ONE WEEK LOAN

Derivatives in Financial Markets
with Stochastic Volatility

This book addresses problems in financial mathematics of pricing and hedging derivative securities in an environment of uncertain and changing market volatility. These problems are important to investors ranging from large trading institutions to pension funds. The authors present mathematical and statistical tools that exploit the "bursty" nature of market volatility. The mathematics is introduced through examples and illustrated with simulations, and the approach described is validated and tested on market data.

The material is suitable for a one-semester course for graduate students who have been exposed to methods of stochastic modeling and arbitrage pricing theory in finance. It is easily accessible to derivatives practitioners in the financial engineering industry.

Jean-Pierre Fouque studied at the University Pierre and Marie Curie in Paris. He was Chargé de Recherche at the French CNRS and Maître de Conférences at the Ecole Polytechnique until 1998. He is now Associate Professor of Mathematics at the North Carolina State University.

George Papanicolaou was Professor of Mathematics at the Courant Institute before coming to Stanford in 1993. He is now Robert Grimmett Professor in the Mathematics department at Stanford.

K. Ronnie Sircar is Assistant Professor of Mathematics at the University of Michigan until the summer of 2000, when he joins the Operations Research and Financial Engineering department at Princeton. He received his Ph.D. from Stanford in 1997.

Derivatives in Financial Markets with Stochastic Volatility

JEAN-PIERRE FOUQUE
North Carolina State University

GEORGE PAPANICOLAOU
Stanford University

K. RONNIE SIRCAR
University of Michigan

CAMBRIDGE
UNIVERSITY PRESS

PUBLISHED BY THE PRESS SYNDICATE OF THE UNIVERSITY OF CAMBRIDGE
The Pitt Building, Trumpington Street, Cambridge, United Kingdom

CAMBRIDGE UNIVERSITY PRESS
The Edinburgh Building, Cambridge CB2 2RU, UK   www.cup.cam.ac.uk
40 West 20th Street, New York, NY 10011-4211, USA   www.cup.org
10 Stamford Road, Oakleigh, Melbourne 3166, Australia
Ruiz de Alarcón 13, 28014 Madrid, Spain

First published 2000

Printed in the United States of America

Typeface Times 10/13 pt.   *System* AMS-T$_E$X   [FH]

A catalog record for this book is available from the British Library

Library of Congress Cataloging in Publication Data
Fouque, Jean-Pierre.
Derivatives in financial markets with stochastic volatility / Jean-Pierre Fouque,
George Papanicolaou, K. Ronnie Sircar.
p.   cm.
Includes bibliographical references.
ISBN 0-521-79163-4
1. Derivative securities.   2. Financial institutions.   I. Papanicolaou, George.
II. Sircar, K. Ronnie (Kaushik Ronnie), 1970–   III. Title.
HG6024.A3   F68   2000
332.63′2 – dc21   00-023603

ISBN 0 521 79163 4 hardback

Contents

Introduction

This book addresses problems in financial mathematics of pricing and hedging derivative securities in an environment of uncertain and changing market volatility. These problems are important to investors ranging from large trading institutions to pension funds. We introduce and systematically present mathematical and statistical tools that we have found to be very effective in this context. The material is suitable for a one-semester course for graduate students who have been exposed to methods of stochastic modeling and arbitrage pricing theory in finance. We have also aimed to make it easily accessible to derivatives practitioners in the financial engineering industry.

It is widely recognized that the simplicity of the popular Black–Scholes model, which relates derivative prices to current stock prices and quantifies risk through a *constant* volatility parameter, is no longer sufficient to capture modern market phenomena – especially since the 1987 crash. The natural extension of the Black–Scholes model that has been pursued in the literature and in practice is to modify the specification of volatility to make it a stochastic process. What makes this approach particularly challenging is first that volatility is a *hidden* process: it is driving prices and yet cannot be directly observed. Second, volatility tends to fluctuate at a high level for a while, then at a low level for a similar period, then high again, and so on. It "mean reverts" many times during the life of a derivative contract.

We describe here a method for modeling, analysis, and estimation that exploits this fast mean reversion of the volatility. It identifies the important groupings of market parameters, which otherwise are not obvious, and it turns out that estimation of these composites from market data is extremely efficient and stable. Furthermore, the methodology is robust in that it does not assume a specific volatility model.

Outline

In Chapter 1 we review the basic ideas and methods of the Black–Scholes theory as well as the stochastic calculus underpinning the models used. Chapter 2 motivates the stochastic volatility models and explains the difficulties induced by them. The main idea of volatility clustering, or *fast mean reversion,* is introduced in Chapter 3, where we describe – from scratch and with examples – the mathematics of this phenomenon. Tools for analyzing data to assess the rate of mean reversion of volatility are presented in Chapter 4. Our analysis shows clearly that the S&P 500 volatility is fast mean-reverting.

The problems we are interested in fall into two broad categories: pricing and hedging. In Chapters 5 and 6 we develop the asymptotic method that exploits volatility clustering for European derivatives. The theory identifies three group parameters that encode the effect of fast mean-reverting market volatility. We describe how these are estimated from the observed implied volatility skew and demonstrate their stability over time in the case of S&P 500 index options.

The extensions to exotic and American claims are described in Chapters 8 and 9, respectively. We outline in Chapter 11 how the tools are effective for fixed-income markets as well.

In Chapter 7 we describe hedging-type problems and how the asymptotic methodology is of use. These problems are somewhat different because, as we explain there, uncertain volatility gives rise to an incomplete market, where the effects of randomly changing volatility cannot be offset or hedged perfectly (in contrast to a known volatility, complete market environment).

Chapter 10 outlines various generalizations of the method: multidimensional situations, periodic effects, and non-Markovian models. We also show how it can be applied in other contexts such as the Merton portfolio optimization problem.

An overview of the problems presented in this book is shown in Figure 1.

Correcting Black–Scholes for Stochastic Volatility

The central idea of the book, presented in Chapters 3–6, is as follows. If volatility were running extremely fast then the market would behave as in a constant volatility Black–Scholes model. We will learn how this *effective volatility* arises. Then, since volatility is running fast but not extremely fast, we can treat the market as a perturbation of a constant volatility Black–Scholes market. We show how to compute the *correction,* which will reflect the effect of stochastic volatility on derivative prices. It involves the third derivative with respect to the price of the classical Black–Scholes price; we call this new "Greek" the *Epsilon.* What is

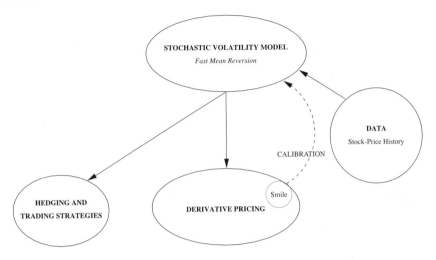

Figure 1. Overview of the problems discussed.

surprising is that all that is needed to compute this correction is the observed price and two quantities that are easily calibrated from the implied volatility surface. This calibration issue is discussed in Chapter 5.

A method for reducing hedging error is presented in Chapter 7. The method can be summarized in the case of European derivatives as follows.

(1) Fit an *affine function* of the log-moneyness-to-maturity ratio (LMMR) to the implied volatility surface across strikes and maturities for liquid European calls:

$$I = a \left(\frac{\log\left(\frac{\text{strike price}}{\text{asset price}} \right)}{\text{time to maturity}} \right) + b.$$

(2) From the estimated slope a and intercept b and the average volatility $\bar{\sigma}$ estimated from historical price data, deduce the two fundamental quantities

$$V_2 = \bar{\sigma}((\bar{\sigma} - b) - a(r + \tfrac{3}{2}\bar{\sigma}^2)),$$

$$V_3 = -a\bar{\sigma}^3,$$

which are small in the range of application of the method. We show that V_3 is directly related to the volatility *skew* and that V_2 contains the market price of volatility *risk*.

(3) Compute the Black–Scholes pricing function $P_0(t, x)$, which gives the price as a function of the present time t and underlying asset price x. For example, one could solve the Black–Scholes partial differential equation with the constant volatility $\bar{\sigma}$ and the appropriate terminal payoff.

(4) When the time to maturity is $T - t$, the *corrected Black–Scholes price* is given by

$$P_0(t, x) - (T - t)\left(V_2 x^2 \frac{\partial^2 P_0}{\partial x^2} + V_3 x^3 \frac{\partial^3 P_0}{\partial x^3}\right).$$

Observe that this computation does not require estimation of the present volatility and is *model-independent* in the sense that we are not calibrating any particular fully specified stochastic volatility model. The correction accounts for the market price of volatility risk.

(5) This approach leads to the following *hedging strategy*: Compute

$$Q = P_0 - (T - t)V_3\left(2x^2 \frac{\partial^2 P_0}{\partial x^2} + x^3 \frac{\partial^3 P_0}{\partial x^3}\right).$$

The proposed hedging strategy consists of holding

$$\frac{\partial Q}{\partial x}(t, x)$$

units of stock and

$$Q(t, x) - x\frac{\partial Q}{\partial x}(t, x)$$

worth of bonds, along the path of the stock-price process. This strategy *replicates* the claim at maturity. It is not self-financing, but we show that the correction compensates a systematic bias and reduces the variance of the cost. We then identify the remaining small nonhedgable part of that cost.

In the case of exotic derivatives treated in Chapter 8 and American derivatives in Chapter 9, we show that the same two quantities V_2 and V_3 are needed to implement this method. Analogous quantities for fixed-income markets are introduced in Chapter 11.

Acknowledgment. K.R.S. acknowledges NSF support, award number DMS-9803169.

1 The Black–Scholes Theory of Derivative Pricing

The aim of this first chapter is to review the basic objects, ideas, and results of the now classical Black–Scholes theory of derivative pricing. It is intended for readers who want to enter the subject or simply refresh their memory. This is not a complete treatment of this theory with detailed proofs but rather an intuitive but precise presentation that includes a few key calculations. Detailed presentations of the subject can be found in many books at various levels of mathematical rigor and generality, a few of which we list in the notes at the end of the chapter.

This book is about correcting the Black–Scholes theory in order to handle markets with stochastic volatility. The notation and many of the tools used in the constant volatility case will be used for the more complex markets throughout the book.

1.1 Market Model

In this simple model, suggested by Samuelson and used by Black and Scholes, there are two assets. One is a riskless asset (bond) with price β_t at time t, described by the ordinary differential equation

$$d\beta_t = r\beta_t \, dt, \tag{1.1}$$

where r, a nonnegative constant, is the instantaneous interest rate for lending or borrowing money. Setting $\beta_0 = 1$, we have $\beta_t = e^{rt}$ for $t \geq 0$. The price X_t of the other asset, the risky stock or stock index, evolves according to the stochastic differential equation

$$dX_t = \mu X_t \, dt + \sigma X_t \, dW_t, \tag{1.2}$$

where μ is a constant mean return rate, $\sigma > 0$ is a constant *volatility*, and $(W_t)_{t \geq 0}$ is a standard Brownian motion. This fundamental model and the intuitive content of equation (1.2) are presented in the following sections.

1.1.1 Brownian Motion

The Brownian motion is a stochastic process whose definition, existence, properties, and applications have been (and still are) the subject of numerous studies during the twentieth century. Our goal here is to give a very intuitive and practical presentation.

Brownian motion is a real-valued stochastic process with continuous trajectories that have independent and stationary increments. The trajectories are denoted by $t \to W_t$ and, for the standard Brownian motion, we have that:

- $W_0 = 0$;
- for any $0 < t_1 < \cdots < t_n$, the random variables $(W_{t_1}, W_{t_2} - W_{t_1}, \ldots, W_{t_n} - W_{t_{n-1}})$ are independent;
- for any $0 \leq s < t$, the increment $W_t - W_s$ is a centered (mean-zero) normal random variable with variance $I\!\!E\{(W_t - W_s)^2\} = t - s$. In particular, W_t is $\mathcal{N}(0, t)$-distributed.

We denote by $(\Omega, \mathcal{F}, I\!\!P)$ the probability space where our Brownian motion is defined and the expectation $I\!\!E\{\cdot\}$ is computed. For instance, it could be $\Omega = \mathcal{C}([0, +\infty) : I\!\!R)$, the space of all continuous trajectories ω such that $W_t(\omega) = \omega(t)$. The σ-algebra $\mathcal{F}$ contains all sets of the form $\{\omega \in \Omega : |\omega(s)| < R, s \leq t\}$; $I\!\!P$ is the Wiener measure, which is the probability distribution of the standard Brownian motion.

The increasing family of σ-algebras $\mathcal{F}_t$ generated by $(W_s)_{s \leq t}$, the information on W up to time t, and all the sets of probability 0 in $\mathcal{F}$ is called the *natural filtration* of the Brownian motion. This *completion* by the null sets is important, in particular for the following reason. If two random variables X and Y are equal almost surely ($X = Y$ $I\!\!P$-a.s. means $I\!\!P\{X = Y\} = 1$) and if X is $\mathcal{F}_t$-measurable (meaning that any event $\{X_t \leq x\}$ belongs to $\mathcal{F}_t$), then Y is also $\mathcal{F}_t$-measurable.

A stochastic process $(X_t)_{t \geq 0}$ is *adapted* to the filtration $(\mathcal{F}_t)_{t \geq 0}$ if the random variable X_t is $\mathcal{F}_t$-measurable for every t. We say that (X_t) is $(\mathcal{F}_t)$-*adapted*. If another process (Y_t) is such that $X_t = Y_t$ $I\!\!P$-a.s. for every t, then it is also $(\mathcal{F}_t)$-adapted.

The independence of the increments of the Brownian motion and their normal distribution can be summarized using *conditional characteristic functions*. For $0 \leq s < t$ and $u \in I\!\!R$,

$$I\!\!E\{e^{iu(W_t - W_s)} \mid \mathcal{F}_s\} = e^{-u^2(t-s)/2}. \tag{1.3}$$

If W is a Brownian motion then, by independence of the increment $W_t - W_s$ from the past $\mathcal{F}_s$, the left-hand side of (1.3) is simply $I\!E\{e^{iu(W_t - W_s)}\}$, which is the characteristic function of a centered normal random variable with variance $t - s$, and is equal to the right-hand side. Conversely, if (1.3) holds then the continuous process (W_t) is a standard Brownian motion.

This independence of increments makes the Brownian motion an ideal candidate to define a complete family of independent infinitesimal increments dW_t, which are centered and normally distributed with variance dt and which will serve as a model of (Gaussian white) noise. The drawback is that the trajectories of (W_t) cannot be "nice" in the sense that they are not of bounded variation, as the following simple computation suggests. Let $t_0 = 0 < t_1 < \cdots < t_n = t$ be a subdivision of $[0, t]$, which we may suppose evenly spaced, so that $t_i - t_{i-1} = t/n$ for each interval. The quantity

$$I\!E\left\{\sum_{i=1}^{n} |W_{t_i} - W_{t_{i-1}}|\right\} = n\,I\!E\{|W_{t/n}|\} = n\sqrt{t/n}\,I\!E\{|W_1|\}$$

goes to $+\infty$ as $n \nearrow +\infty$, indicating that the integral with respect to dW_t cannot be defined in the usual way "trajectory by trajectory." We describe how such integrals can be defined in the next section.

1.1.2 Stochastic Integrals

For T a fixed finite time, let $(X_t)_{0 \le t \le T}$ be a stochastic process adapted to $(\mathcal{F}_t)_{0 \le t \le T}$, the filtration of the Brownian motion up to time T, such that

$$I\!E\left\{\int_0^T (X_t)^2\, dt\right\} < +\infty. \tag{1.4}$$

Using iterated conditional expectations and the independent increments property of Brownian motion, we note that

$$I\!E\left\{\left(\sum_{i=1}^{n} X_{t_{i-1}}(W_{t_i} - W_{t_{i-1}})\right)^2\right\} = I\!E\left\{\sum_{i=1}^{n} (X_{t_{i-1}})^2(t_i - t_{i-1})\right\}$$

for $t \le T$, which is a basic calculation in the construction of stochastic integrals. Note also that the Brownian increments on the left are forward in time and that the sum on the right converges to $I\!E\{\int_0^t (X_s)^2\, ds\}$, which, by (1.4), is finite.

The *stochastic integral* of (X_t) with respect to the Brownian motion (W_t) is defined as a limit in the mean-square sense ($L^2(\Omega)$),

$$\int_0^t X_s \, dW_s = \lim_{n \nearrow +\infty} \sum_{i=1}^n X_{t_{i-1}}(W_{t_i} - W_{t_{i-1}}), \tag{1.5}$$

as the mesh size of the subdivision goes to zero.

As a function of time t, this stochastic integral defines a continuous square integrable process such that

$$I\!E\left\{ \left(\int_0^t X_s \, dW_s \right)^2 \right\} = I\!E\left\{ \int_0^t X_s^2 \, ds \right\}. \tag{1.6}$$

It has the *martingale property*

$$I\!E\left\{ \int_0^t X_u \, dW_u \mid \mathcal{F}_s \right\} = \int_0^s X_u \, dW_u \quad (I\!P\text{-a.s.}, \ s \le t), \tag{1.7}$$

as can be easily deduced from the definition (1.5). The *quadratic variation* $\langle Y \rangle_t$ of the stochastic integral $Y_t = \int_0^t X_u \, dW_u$ is

$$\langle Y \rangle_t = \lim_{n \nearrow +\infty} \sum_{i=1}^n (Y_{t_i} - Y_{t_{i-1}})^2 = \int_0^t X_s^2 \, ds \tag{1.8}$$

in the mean-square sense.

Stochastic integrals are mean-zero, continuous, and square integrable martingales. It is interesting to note that the converse is also true: every mean-zero, continuous, and square integrable martingale is a Brownian stochastic integral. This will be made precise and used in Section 1.4.

1.1.3 Risky Asset Price Model

The Black–Scholes model for the risky asset price corresponds to a continuous process (X_t) such that, in an infinitesimal amount of time dt, the infinitesimal return dX_t/X_t has mean $\mu \, dt$, proportional to dt, with a constant *rate of return* μ and centered random fluctuations that are independent of the past up to time t. These fluctuations are modeled by $\sigma \, dW_t$ where σ is a positive constant *volatility* and dW_t the infinitesimal increments of the Brownian motion. The corresponding formula for the infinitesimal return is

$$\frac{dX_t}{X_t} = \mu \, dt + \sigma \, dW_t, \tag{1.9}$$

which is the stochastic differential equation (1.2). The right side has the natural financial interpretation of a return term plus a risk term. We are also assuming that there are no dividends paid in the time interval that we are considering. It is easy to incorporate a continuous dividend rate in all that follows, but for simplicity we shall omit this.

In integral form, this equation is

$$X_t = X_0 + \mu \int_0^t X_s \, ds + \sigma \int_0^t X_s \, dW_s, \tag{1.10}$$

where the last integral is a stochastic integral as described in Section 1.1.2 and where X_0 is the initial value, which is assumed to be independent of the Brownian motion and square integrable.

Equation (1.10), or (1.2) in the differential form, is a particular case of a general class of stochastic differential equations driven by a Brownian motion:

$$dX_t = \mu(t, X_t) \, dt + \sigma(t, X_t) \, dW_t \tag{1.11}$$

or, in integral form,

$$X_t = X_0 + \int_0^t \mu(s, X_s) \, ds + \int_0^t \sigma(s, X_s) \, dW_s. \tag{1.12}$$

In the Black–Scholes model, $\mu(t, x) = \mu x$ and $\sigma(t, x) = \sigma x$; these are independent of t, differentiable in x, and linearly growing at infinity (since they are linear). This is enough to ensure existence and uniqueness of a continuous adapted and square integrable solution (X_t). The proof of this result is based on simple estimates like

$$\mathbb{E}\{X_t^2\} = \mathbb{E}\left\{\left(X_0 + \mu \int_0^t X_s \, ds + \sigma \int_0^t X_s \, dW_s\right)^2\right\}$$

$$\leq 3\left(\mathbb{E}\{X_0^2\} + (\mu^2 T + \sigma^2) \int_0^t \mathbb{E}\{X_s^2\} \, ds\right),$$

where we have used the inequality $(a + b + c)^2 \leq 3(a^2 + b^2 + c^2)$, the Cauchy–Schwarz inequality

$$\mathbb{E}\left(\int_0^t X_s \, ds\right)^2 \leq t \int_0^t \mathbb{E}\{X_s^2\} \, ds,$$

and (1.6). We deduce that

$$0 \leq \mathbb{E}\{X_t^2\} \leq c_1 + c_2 \int_0^t \mathbb{E}\{X_s^2\} \, ds$$

for $0 \leq t \leq T$ and constants $c_1, c_2 \geq 0$. By a direct application of Gronwall's lemma, we deduce that the solution is a priori square integrable. The construction of a solution and the proof of uniqueness can be obtained by similar and slightly more complicated estimates that use the Kolmogorov–Doob inequality for martingales.

Looking at equation (1.9), it is very tempting to write X_t/X_0 explicitly as the exponential of $(\mu t + \sigma W_t)$. However, this is not correct because the usual chain rule is not valid for stochastic differentials. For instance, W_t^2 is not equal to $2 \int_0^t W_s \, dW_s$ as might be expected since, by the martingale property (1.7), this last integral has an expectation equal to zero but $I\!E\{W_t^2\} = t$.

   This discrepancy is corrected by Itô's formula, as explained in the following section.

1.1.4 Itô's Formula

A function of the Brownian motion W_t defines a new stochastic process $g(W_t)$. We suppose in the following that the function g is twice continuously differentiable, bounded, and has bounded derivatives. The purpose of the chain rule is to compute the differential $dg(W_t)$ or equivalently its integral $g(W_t) - g(W_0)$. Using the subdivision $t_0 = 0 < t_1 \cdots < t_n = t$ of $[0, t]$, we write

$$g(W_t) - g(W_0) = \sum_{i=1}^{n} (g(W_{t_i}) - g(W_{t_{i-1}})).$$

We then apply Taylor's formula to each term to obtain

$$g(W_t) - g(W_0) = \sum_{i=1}^{n} g'(W_{t_{i-1}})(W_{t_i} - W_{t_{i-1}})$$

$$+ \frac{1}{2} \sum_{i=1}^{n} g''(W_{t_{i-1}})(W_{t_i} - W_{t_{i-1}})^2 + R,$$

where R contains all the higher-order terms.

   If (W_t) were differentiable only the first sum would contribute to the limit as the mesh size of the subdivision goes to zero, leading to the chain rule $dg(W_t) = g'(W_t)W_t' \, dt$ of calculus. In the Brownian case (W_t) is not differentiable and, by (1.5), the first sum converges to the stochastic integral

$$\int_0^t g'(W_s) \, dW_s.$$

The correction comes from the second sum, which, like (1.8), converges to

$$\frac{1}{2} \int_0^t g''(W_s) \, ds;$$

this can be seen by comparing it in L^2 with $\frac{1}{2} \sum_{i=1}^{n} g''(W_{t_{i-1}})(t_i - t_{i-1})$. The higher-order terms contained in R converge to zero and do not contribute to the limit, which is

$$g(W_t) - g(W_0) = \int_0^t g'(W_s) \, dW_s + \frac{1}{2} \int_0^t g''(W_s) \, ds. \qquad (1.13)$$

This is the simplest version of Itô's formula. It is often written in differential form:

$$dg(W_t) = g'(W_t)\, dW_t + \tfrac{1}{2} g''(W_t)\, dt. \qquad (1.14)$$

The next step is deriving a similar formula for $dg(X_t)$, where X_t is the solution of a stochastic differential equation like (1.11). We give here this general formula for a function g depending also on time t:

$$dg(t, X_t) = \frac{\partial g}{\partial t}(t, X_t)\, dt + \frac{\partial g}{\partial x}(t, X_t)\, dX_t + \frac{1}{2}\frac{\partial^2 g}{\partial x^2}(t, X_t)\, d\langle X \rangle_t, \qquad (1.15)$$

where dX_t is given by the stochastic differential equation (1.11) and $\langle X \rangle_t = \int_0^t \sigma^2(s, X_s)\, ds$ is the quadratic variation of the martingale part of X_t, that is, the stochastic integral on the right side of (1.12). In terms of dt and dW_t, the formula is

$$dg(t, X_t) = \left(\frac{\partial g}{\partial t} + \mu(t, X_t)\frac{\partial g}{\partial x} + \frac{1}{2}\sigma^2(t, X_t)\frac{\partial^2 g}{\partial x^2} \right) dt + \sigma(t, X_t)\frac{\partial g}{\partial x}\, dW_t, \qquad (1.16)$$

where all the partial derivatives of g are evaluated at (t, X_t).

As an application we can compute the differential of the discounted price $g(t, X_t) = e^{-rt} X_t$:

$$d(e^{-rt} X_t) = -re^{-rt} X_t\, dt + e^{-rt}\, dX_t$$

$$= e^{-rt}(-rX_t + \mu(t, X_t))\, dt + e^{-rt}\sigma(t, X_t)\, dW_t, \qquad (1.17)$$

since the second derivative of $g(t, x) = xe^{-rt}$ with respect to x is zero. In the particular case of the price X_t given by (1.2), $\mu(t, x) = \mu x$ and $\sigma(t, x) = \sigma x$ so we obtain

$$d(e^{-rt} X_t) = (\mu - r)(e^{-rt} X_t)\, dt + \sigma(e^{-rt} X_t)\, dW_t. \qquad (1.18)$$

The discounted price $\widetilde{X}_t = e^{-rt} X_t$ satisfies the same equation as X_t, where the return μ has been replaced by $\mu - r$.

The previous example is a particular case of the useful general *integration by parts* formula, which consists of computing the differential of the product of two processes that are solutions of equations like (1.11). If X and Y denote these processes and if $\mu_X, \sigma_X, \mu_Y, \sigma_Y$ denote the coefficients of their corresponding equations, then this formula is

$$d(X_t Y_t) = X_t\, dY_t + Y_t\, dX_t + d\langle X, Y \rangle_t, \qquad (1.19)$$

where the covariation (also called the "bracket") of X and Y is given by

$$d\langle X, Y \rangle_t = \sigma_X(t, X_t)\sigma_Y(t, Y_t)\, dt.$$

This follows from the identity

$$XY = \tfrac{1}{4}((X + Y)^2 - (X - Y)^2)$$

and applying Itô's lemma to $(X_t \pm Y_t)^2$.

Note that $\langle X, X \rangle_t$ is simply $\langle X \rangle_t$, the quadratic variation (1.8) of the martingale part of (X_t).

1.1.5   Lognormal Risky Asset Price

Coming back to the stochastic differential equation (1.9) for the evolution of the stock price X_t, it is natural to suspect from the ordinary calculus formula $\int dx/x = \log x$ that $\log X_t$ might satisfy an equation that we can integrate explicitly. We compute the differential of $\log X_t$ by applying Itô's formula (1.16) with $g(t, x) = \log x$, $\mu(t, x) = \mu x$, and $\sigma(t, x) = \sigma x$:

$$d \log X_t = (\mu - \tfrac{1}{2}\sigma^2)\, dt + \sigma\, dW_t.$$

The logarithm of the stock price is then given explicitly by

$$\log X_t = \log X_0 + (\mu - \tfrac{1}{2}\sigma^2)t + \sigma W_t,$$

which leads to the following formula for the stock price:

$$X_t = X_0 \exp((\mu - \tfrac{1}{2}\sigma^2)t + \sigma W_t). \tag{1.20}$$

The return X_t/X_0 is *lognormal*: it is the exponential of a nonstandard Brownian motion that is normally distributed with mean $(\mu - \tfrac{1}{2}\sigma^2)t$ and variance $\sigma^2 t$ at time t. The process (X_t) is also called *geometric* Brownian motion. The stock price given by (1.20) satisfies equation (1.9). It can also be obtained as a diffusion limit of binomial tree models, which arise when Brownian motion is approximated by a random walk.

Notice that, if X_t becomes zero, it stays at zero for all times thereafter. Thus, in this model, bankruptcy (zero stock price) is a permanent state. However, $\tfrac{1}{t}W_t$ tends to zero as t tends to infinity with probability 1, so it follows that if X_0 is not zero then (with probability 1) X_t does not go to zero in a finite time.

In Figure 1.1, we show a sample path or realization of a geometric Brownian motion (X_t) in which $\mu = 0.15$, $\sigma = 0.1$, and $X_0 = 95$. This path exhibits the "average growth plus noise" behavior we expect from this model of asset prices.

1.2   Derivative Contracts

Derivatives are contracts based on the underlying asset price (X_t). They are also called *contingent claims*. We will be interested primarily in *options*, which can be

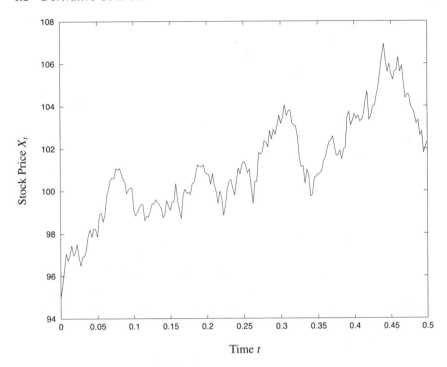

Figure 1.1. A sample path of a geometric Brownian motion defined by the stochastic differential equation (1.9), with $\mu = 0.15$, $\sigma = 0.1$, and $X_0 = 95$.

European, American, path-independent or path-dependent. The definition of the options discussed in this chapter is given in the following sections.

1.2.1   *European Call and Put Options*

A *European call option* is a contract that gives its holder the right, but not the obligation, to buy one unit of an underlying asset for a predetermined *strike price K* on the *maturity* date T. If X_T is the price of the underlying asset at maturity time T, then the value of this contract at maturity, its *payoff*, is

$$h(X_T) = (X_T - K)^+ = \begin{cases} X_T - K & \text{if } X_T > K, \\ 0 & \text{if } X_T \le K, \end{cases} \tag{1.21}$$

since in the first case the holder will exercise the option and make a profit $X_T - K$ by buying the stock for K and selling it immediately at the market price X_T. In the second case the option is not exercised, since the market price of the asset is less than the strike price.

Similarly, a *European put option* is a contract that gives its holder the right to sell a unit of the asset for a strike price K at the maturity date T. Its payoff is

$$h(X_T) = (K - X_T)^+ = \begin{cases} K - X_T & \text{if } X_T < K, \\ 0 & \text{if } X_T \geq K. \end{cases} \tag{1.22}$$

In the first case, buying the stock at the market price and exercising the put option yields a profit of $K - X_T$; in the second case, the option is simply not exercised.

More generally, we will consider European derivatives defined by their maturity time T and their nonnegative payoff function $h(x)$. This will be a contract that pays $h(X_T)$ at maturity time T when the stock price is X_T. The standard European-style derivatives are *path-independent* because the payoff $h(X_T)$ is only a function of the value of the stock price at maturity time T.

At time $t < T$ this contract has a value, known as the *derivative price,* which will vary with t and the observed stock price X_t. This option price at time t for a stock price $X_t = x$ is denoted by $P(t, x)$, and the problem of *derivative pricing* is to determine this pricing function. The fact that this option price will depend only on the observed value at time t and not on the past values of the stock price is closely related to the *Markov property* shared by the solutions of stochastic differential equations like (1.11), by which we shall model the stock price. More details on this will be given in Section 1.5.

Perhaps the simplest way to price such a derivative is as the expected value of the discounted payoff. More precisely, if the stock price is the process (1.2) and if the observed stock price is $X_0 = x$, then the option price at time $t = 0$ is

$$P(0, x) = \mathbb{E}\{e^{-rT} h(X_T)\}$$
$$= \mathbb{E}\{e^{-rT} h(x e^{(\mu - \sigma^2/2)T + \sigma W_T})\}, \tag{1.23}$$

where we have used the explicit formula (1.20) with $X_0 = x$ and time T. The expectation reduces to a Gaussian integral, since W_T is $\mathcal{N}(0, T)$-distributed. In general (unless $\mu = r$), the option price given by formula (1.23) leads to an *arbitrage opportunity,* meaning that there will be a risk-free way to make a profit with strictly positive expectation by holding a particular portfolio. This is one of the key ideas in Section 1.3 that is used to determine the *fair* option price.

1.2.2 American Options

An *American option* is a contract in which the holder decides whether to exercise the option or not at any time of his choice before the option's expiration date T. The time τ at which the option is exercised is called the *exercise time.* Because the market cannot be anticipated, the holder of the option has to decide to exercise or not at time $t \leq T$ with information up to time t contained in the σ-algebra $\mathcal{F}_t$. In other words, τ is a random time such that the event $\{\tau \leq t\}$ (or its complement $\{\tau > t\}$) belongs to $\mathcal{F}_t$ for any $t \leq T$. Such a random time is called a *stopping time* with respect to the filtration $(\mathcal{F}_t)$. If the payoff function of the derivative is h then its value at the exercise time τ is $h(X_\tau)$, where X_τ is the stock price at the stopping time τ.

For an *American call option* the payoff is $h(X_\tau) = (X_\tau - K)^+$ for a given strike price K and a stopping time $\tau \leq T$ chosen by the holder of the option. Note that even if τ is an exercise time the option will be exercised only if $X_\tau > K$ but, in any case, the contract is terminated at time τ. Similarly, the payoff of an *American put option* is $h(X_\tau) = (K - X_\tau)^+$ and the option is exercised only if $K > X_\tau$.

As in the case of European derivatives, an intuitive way to price an American derivative (at time $t = 0$) is to maximize the expected value of the discounted payoff over all the stopping times $\tau \leq T$:

$$P(0, x) = \sup_{\tau \leq T} I\!E\{e^{-r\tau} h(X_\tau)\}. \tag{1.24}$$

Again, this price leads in general to an opportunity for arbitrage and therefore cannot be the fair price of the derivative.

1.2.3 Other Exotic Options

The term "exotic option" refers here to any option contract that is not a standard European or American option described in the previous sections. Our aim is not to write a catalog of existing options but rather to give some examples of exotic options that we will use in the rest of the book.

Barrier options are path-dependent options whose payoff depends on whether or not the underlying asset price hits a specified value during the option's lifetime. For instance, a *down-and-out call option* becomes worthless (or "knocked out") if, at any time t before the expiration date T, the stock price X_t falls below a predetermined level B. The payoff at expiration T is a function of the trajectory of the stock price

$$h = (X_T - K)^+ \mathbf{1}_{\{\inf_{t \leq T} X_t > B\}}. \tag{1.25}$$

Here $\mathbf{1}_A(x) = 1$ if $x \in A$ and $\mathbf{1}_A(x) = 0$ if x is not in A. It is the indicator function of the set A. This option is obviously less valuable than a standard European call option given by (1.21) with the same strike K and maturity T, and it will lead to a *knock-out discount*.

Our last example is **Asian options**, whose payoff depends on the average stock price during a specified period of time before maturity. They can be European or American, with typical payoffs like

$$h = \left(X_T - \frac{1}{T} \int_0^T X_s \, ds \right)^+ \tag{1.26}$$

for an *arithmetic-average strike call option* (European style), where the strike price is the average stock price.

1.3   Replicating Strategies

The Black–Scholes analysis of a European-style derivative yields an explicit trading strategy in the underlying risky asset and riskless bond whose terminal payoff is equal to the payoff $h(X_T)$ of the derivative at maturity, no matter what path the stock price takes. Thus, selling the derivative and holding a dynamically adjusted portfolio according to this strategy "covers" an investor against all risk of eventual loss, because a loss incurred at the final time from one part of this portfolio will be exactly compensated by a gain in the other part. This *replicating strategy,* as it is known, therefore provides an insurance policy against the risk of being short the derivative. It is called a *dynamic hedging strategy* since it involves continuous trading, where to hedge means to eliminate risk. The essential step in the Black–Scholes methodology is the construction of this replicating strategy and arguing, based on *no arbitrage*, that the value of the replicating portfolio at time t is the fair price of the derivative. We develop this argument in the following sections.

1.3.1   Replicating Self-Financing Portfolios

We consider a European-style derivative with payoff $h(X_T)$, a function of the underlying asset price at maturity time T. Assume that the stock price (X_t) follows the geometric Brownian motion model (1.20), a solution of the stochastic differential equation (1.2). A *trading strategy* is a pair (a_t, b_t) of adapted processes specifying the number of units held at time t of the underlying asset and the riskless bond, respectively. We suppose that $I\!E\{\int_0^T (a_t)^2 \, dt\}$ and $\int_0^T |b_t| \, dt$ are finite so that the stochastic integral involving (a_t) and the usual integral involving (b_t) are well-defined.

Assuming, as in (1.1), that the price of the bond at time t is $\beta_t = e^{rt}$, the value at time t of this portfolio is $a_t X_t + b_t e^{rt}$. It will *replicate* the derivative at maturity if its value at time T is almost surely equal to the payoff:

$$a_T X_T + b_T e^{rT} = h(X_T). \tag{1.27}$$

In addition, this portfolio is to be *self-financing,* meaning that the variations of its value are due only to the variations of the market – that is, the variations of the stock and bond prices. No further funds are required after the initial investment, so that if, for example, more of the asset is bought (a_t is increased) then money would have to be obtained by selling bonds (b_t decreased) to pay for it. This is expressed in differential form as

$$d(a_t X_t + b_t e^{rt}) = a_t \, dX_t + rb_t e^{rt} \, dt, \tag{1.28}$$

which implies that

$$X_t \, da_t + e^{rt} \, db_t + d\langle a, X\rangle_t = 0, \tag{1.29}$$

using the integration-by-parts formula (1.19) to compute $d(a_t X_t)$. In integral form, the self-financing property is

$$a_t X_t + b_t e^{rt} = a_0 X_0 + b_0 + \int_0^t a_s \, dX_s + \int_0^t rb_s e^{rs} \, ds, \quad 0 \le t \le T.$$

An intuitive way to understand this relation is to think in terms of discrete trading times $\{t_n, \, n = 0, 1, \dots\}$. The portfolio consists of a_{t_n} and b_{t_n} of the stock and bond (respectively) at time t_n. When the prices change to $X_{t_{n+1}}$ and $e^{rt_{n+1}}$, we observe the change and *then* we adjust our holdings to $a_{t_{n+1}}$ and $b_{t_{n+1}}$. As no further cash input or output is allowed, the value of the portfolio after adjustment must equal the value before, so

$$a_{t_n} X_{t_{n+1}} + b_{t_n} e^{rt_{n+1}} = a_{t_{n+1}} X_{t_{n+1}} + b_{t_{n+1}} e^{rt_{n+1}}.$$

This says that

$$a_{t_{n+1}} X_{t_{n+1}} + b_{t_{n+1}} e^{rt_{n+1}} - (a_{t_n} X_{t_n} + b_n e^{rt_n}) = a_{t_n}(X_{t_{n+1}} - X_{t_n}) + b_{t_n}(e^{rt_{n+1}} - e^{rt_n}),$$

which in continuous time becomes (1.28).

1.3.2 The Black–Scholes Partial Differential Equation

As in Section 1.2.1, the pricing function for a European-style contract with payoff $h(X_T)$ is denoted by $P(t, x)$. At this stage we do not even know that we can find such a function relating the option price to the present risky asset price and not to its history. Nevertheless, we shall assume that such a pricing function $P(t, x)$ exists and is regular enough to apply Itô's formula (1.16). Our goal is to construct a self-financing portfolio (a_t, b_t) that will replicate the derivative at maturity (1.27).

Excluding the possibility of arbitrage opportunities requires that

$$a_t X_t + b_t e^{rt} = P(t, X_t) \quad \text{for any } 0 \le t \le T. \tag{1.30}$$

For if at some time $t < T$ the left-hand side of (1.30) is (say) less than the right-hand side, an arbitrage opportunity exists by selling the overpriced derivative security immediately and investing in the underpriced asset–bond trading strategy. This yields an instant profit with no exposure to future loss, since the terminal payoff of the trading strategy is equal to the payoff of the derivative.

Differentiating (1.30) and using the self-financing property (1.28) on the left-hand side, Itô's formula (1.16) on the right-hand side, and equation (1.2), we obtain

$$(a_t \mu X_t + b_t r e^{rt}) \, dt + a_t \sigma X_t \, dW_t = \left(\frac{\partial P}{\partial t} + \mu X_t \frac{\partial P}{\partial x} + \frac{1}{2} \sigma^2 X_t^2 \frac{\partial^2 P}{\partial x^2} \right) dt$$

$$+ \sigma X_t \frac{\partial P}{\partial x} \, dW_t, \tag{1.31}$$

where all the partial derivatives of P are evaluated at (t, X_t). Equating the coefficients of the dW_t terms gives

$$a_t = \frac{\partial P}{\partial x}(t, X_t). \tag{1.32}$$

From (1.30) we obtain

$$b_t = (P(t, X_t) - a_t X_t) e^{-rt}. \tag{1.33}$$

Equating the dt terms in (1.31) gives

$$r \left(P - X_t \frac{\partial P}{\partial x} \right) = \frac{\partial P}{\partial t} + \frac{1}{2} \sigma^2 X_t^2 \frac{\partial^2 P}{\partial x^2}, \tag{1.34}$$

which is satisfied for any stock price X_t if $P(t, x)$ is the solution of the *Black–Scholes partial differential equation*

$$\mathcal{L}_{BS}(\sigma) P = 0, \tag{1.35}$$

where

$$\mathcal{L}_{BS}(\sigma) = \frac{\partial}{\partial t} + \frac{1}{2} \sigma^2 x^2 \frac{\partial^2}{\partial x^2} + r \left(x \frac{\partial}{\partial x} - \cdot \right). \tag{1.36}$$

This equation holds in the domain $t \leq T$ and $x > 0$, since in our model the stock price remains positive. It is to be solved *backward in time* with the final condition $P(T, x) = h(x)$, because at expiration the price of the derivative is simply its payoff.

The partial differential equation (1.35) with its final condition has a unique solution $P(t, x)$, which is the value of a self-financing replicating portfolio. Knowing P, the portfolio (a_t, b_t) is uniquely determined by (1.32) and (1.33).

Surprisingly, the rate of return μ does not enter at all in the valuation of this portfolio, owing to the cancellation of the μ terms in (1.31) after the determination (1.32) of a_t. This is a key fact in setting the fair price of the derivative as an expected payoff (see Section 1.4). It is a remarkable feature of the Black–Scholes theory that if two investors have different speculative views about the growth rate of the risky asset – meaning that they have different values of μ but agree that the (commonly estimated and stable) historical volatility σ will prevail – then they will agree on the no-arbitrage price of the derivative, since P does not depend on μ.

1.3.3 Pricing to Hedge

There is another way to derive the Black–Scholes partial differential equation that emphasizes risk elimination or hedging. It is a reinterpretation of the calculations in the previous section as follows.

Let $P_t = P(t, X_t)$ be the price of the option. If we sell N_t options and hold A_t units of the risky asset X_t, then the change in the value of this portfolio is $A_t\, dX_t - N_t\, dP_t$ because it is assumed to be self-financing. We now determine (A, N) so that this portfolio is riskless, which means that we set the coefficient of dW_t to zero. The change in the value of the portfolio should then equal that of a riskless asset, so

$$A_t\, dX_t - N_t\, dP_t = r(A_t X_t - N_t P_t)\, dt.$$

Using (1.9) and Itô's formula, we have

$$A_t(\mu X_t\, dt + \sigma X_t\, dW_t) - N_t \left\{ \left(\frac{\partial P}{\partial t} + \mu X_t \frac{\partial P}{\partial x} + \frac{1}{2}\sigma^2 X_t^2 \frac{\partial^2 P}{\partial x^2} \right) dt - \sigma X_t \frac{\partial P}{\partial x}\, dW_t \right\}$$
$$= r(A_t X_t - N_t P_t)\, dt.$$

Eliminating the dW_t terms gives

$$A_t = N_t \frac{\partial P}{\partial x}(t, X_t),$$

and then the terms involving μ cancel also. We are thus left with the Black–Scholes partial differential equation (1.35) for $P(t, x)$.

In this derivation of the Black–Scholes pricing equation, the role of hedging is clear. Selling the option and holding a dynamically adjusted amount of the risky asset so as to eliminate risk determines the price of the option P_t and the hedge ratio A_t/N_t. This is known as **Delta hedging**, and the ratio $a_t = A_t/N_t$ given by (1.32) is called the Delta.

1.3.4 The Black–Scholes Formula

For European call options described in Section 1.2.1, the Black–Scholes partial differential equation (1.35) is solved with the final condition $h(x) = (x - K)^+$. Prices of European calls at time t and for an observed risky asset price $X_t = x$ will be denoted by $C_{BS}(t, x)$. In this particular case, there is a closed-form solution known as the *Black–Scholes formula*:

$$C_{BS}(t, x) = x N(d_1) - K e^{-r(T-t)} N(d_2), \tag{1.37}$$

where

$$d_1 = \frac{\log(x/K) + (r + \frac{1}{2}\sigma^2)(T - t)}{\sigma\sqrt{T - t}}, \tag{1.38}$$

$$d_2 = d_1 - \sigma\sqrt{T - t}, \tag{1.39}$$

and

$$N(z) = \frac{1}{\sqrt{2\pi}} \int_{-\infty}^{z} e^{-y^2/2} \, dy. \tag{1.40}$$

This convenient formula for the price of a call option – in terms of the current stock price x, the *time to maturity* $T - t$, the strike price K, the volatility σ, and the interest rate r – explains the popularity of the model in the financial services industry since the mid-1970s. We will also denote C_{BS} by $C_{BS}(t, x; K, T; \sigma)$ to emphasize the dependence on K, T, and σ. Only the volatility σ, the standard deviation of the returns scaled by the square root of the time increment, needs to be estimated from data, assuming that the interest rate r is known.

The fact that $C_{BS}(t, x)$ given by (1.37) satisfies equation (1.35) with the final condition $h(x) = (x - K)^+$ can easily be checked directly. A probabilistic representation of this solution is presented in the following section.

Figure 1.2 shows the pricing function $C_{BS}(0, x; 100, 0.5; 0.1)$ plotted against the present $(t = 0)$ stock price x. Notice how it is a smoothed version of the "ramp" terminal payoff function.

The Delta hedging ratio a_t for a call is given by

$$\frac{\partial C_{BS}}{\partial x}(t, x) = N(d_1).$$

There is a similar formula for European put options. Let $P_{BS}(t, x)$ be the price of a European put option (Section 1.2.1). We then have the *put–call parity* relation

$$C_{BS}(t, X_t) - P_{BS}(t, X_t) = X_t - K e^{-r(T-t)}, \tag{1.41}$$

between put and call options with the same maturity and strike price. This is a model-free relationship that follows from simple no-arbitrage arguments. If the left side is smaller than the right side then buying a call and selling a put and one unit of the stock, and investing the difference in the bond, creates a profit at time T no matter what the stock price is.

This relationship is preserved under the lognormal model because the difference $C_{BS} - P_{BS}$ satisfies the partial differential equation (1.35) with the final condition

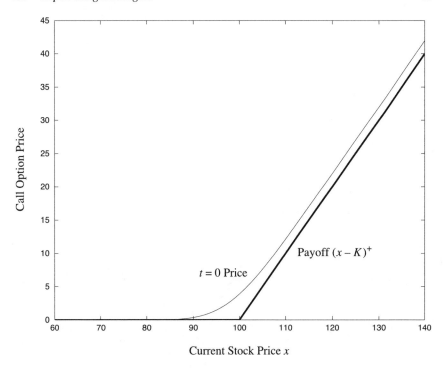

Figure 1.2. Black–Scholes call option pricing function at time $t = 0$, with $K = 100$, $T = 0.5$, $\sigma = 0.1$, and $r = 0.04$.

$h(x) = x - K$. This problem has the simple solution $x - Ke^{-r(T-t)}$. Using the Black–Scholes formula (1.37) for C_{BS} and the put–call parity relation (1.41), we deduce the following explicit formula for the price of a European put option:

$$P_{BS}(t, x) = Ke^{-r(T-t)} N(-d_2) - xN(-d_1), \qquad (1.42)$$

where d_1, d_2, and N are as in (1.38), (1.39), and (1.40), respectively.

Figure 1.3 shows the pricing function $P_{BS}(0, x; 100, 0.5; 0.1)$ plotted against the present ($t = 0$) stock price x. Here we see that the pricing function crosses over its terminal payoff for some (small enough) x, which does not happen with the call option function in Figure 1.2. This observation will be important when we look at American options in Section 1.5.4.

Other types of options do not, in general, lead to such explicit formulas. Determining their prices requires solving numerically the partial differential equation (1.35) with appropriate boundary conditions. Nevertheless, probabilistic representations can be derived, as explained in the following section. In particular,

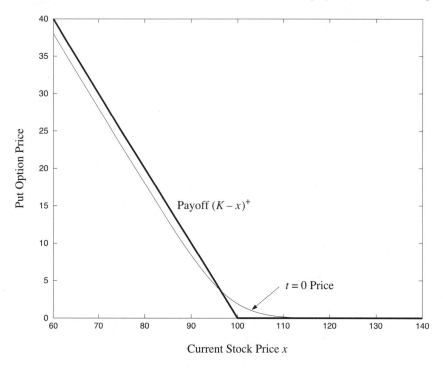

Figure 1.3. Black–Scholes put option pricing function at time $t = 0$, with $K = 100$, $T = 0.5$, $\sigma = 0.1$, and $r = 0.04$.

American options lead to *free boundary value problems* associated with equation (1.35).

1.4 Risk-Neutral Pricing

We mentioned in Section 1.2.1 that, unless $\mu = r$, the expected value under the *subjective* probability $I\!P$ of the discounted payoff of a derivative (1.23) would lead to an opportunity for arbitrage. This is closely related to the fact that the discounted price $\widetilde{X}_t = e^{-rt}X_t$ is not a martingale since, from (1.18),

$$d\widetilde{X}_t = (\mu - r)\widetilde{X}_t \, dt + \sigma \widetilde{X}_t \, dW_t, \tag{1.43}$$

which contains a nonzero *drift* term if $\mu \neq r$.

The main result we want to build in this section is that there is a unique probability measure $I\!P^\star$ equivalent to $I\!P$ such that, under this probability, (i) the discounted price $\widetilde{X}_t$ is a martingale and (ii) the expected value under $I\!P^\star$ of the discounted payoff of a derivative gives its no-arbitrage price. Such a probability measure describing a *risk-neutral* world is called an *equivalent martingale measure*.

1.4.1 Equivalent Martingale Measure

In order to find a probability measure under which the discounted price $\widetilde{X}_t$ is a martingale, we rewrite (1.43) in such a way that the drift term is "absorbed" into the martingale term:

$$d\widetilde{X}_t = \sigma \widetilde{X}_t \left[dW_t + \left(\frac{\mu - r}{\sigma} \right) dt \right].$$

We set

$$\theta = \frac{\mu - r}{\sigma}, \tag{1.44}$$

called the *market price of asset risk*, and define

$$W_t^\star = W_t + \int_0^t \theta \, ds = W_t + \theta t, \tag{1.45}$$

so that

$$d\widetilde{X}_t = \sigma \widetilde{X}_t \, dW_t^\star. \tag{1.46}$$

Using the characterization (1.3), it is easy to check that the positive random variable ξ_T^θ defined by

$$\xi_T^\theta = \exp(-\theta W_T - \tfrac{1}{2}\theta^2 T) \tag{1.47}$$

has an expected value (with respect to $I\!P$) equal to 1 (the Cameron–Martin formula). More generally, it has a conditional expectation with respect to $\mathcal{F}_t$ given by

$$I\!E\{\xi_T^\theta \mid \mathcal{F}_t\} = \exp(-\theta W_t - \tfrac{1}{2}\theta^2 t) = \xi_t^\theta \quad \text{for } 0 \leq t \leq T,$$

which defines a martingale denoted by $(\xi_t^\theta)_{0 \leq t \leq T}$.

We now introduce the probability measure $I\!P^\star$. It is an equivalent measure to $I\!P$, meaning that it has the same null sets ($I\!P^\star$ and $I\!P$ agree on which events have zero probability); $I\!P^\star$ has the density ξ_T^θ with respect to $I\!P$,

$$dI\!P^\star = \xi_T^\theta \, dI\!P. \tag{1.48}$$

Denoting by $I\!E^\star\{\cdot\}$ the expectation with respect to $I\!P^\star$, for any integrable random variable Z we have

$$I\!E^\star\{Z\} = I\!E\{\xi_T^\theta Z\};$$

one can check that, for any adapted and integrable process (Z_t),

$$I\!E^\star\{Z_t \mid \mathcal{F}_s\} = \frac{1}{\xi_s^\theta} I\!E\{\xi_t^\theta Z_t \mid \mathcal{F}_s\} \tag{1.49}$$

for any $0 \le s \le t \le T$. The process $(\xi_t^\theta)_{0 \le t \le T}$ is called the Radon–Nikodym process.

The main result of this section asserts that the process (W_t^*) given by (1.45) is a standard Brownian motion under the probability $I\!P^*$. This result in its full generality (when θ is an adapted stochastic process) is known as *Girsanov's theorem*. In our simple case (θ constant), it is easily derived by using the characterization (1.3) and formula (1.49) as follows:

$$
\begin{aligned}
I\!E^*\{e^{iu(W_t^* - W_s^*)} \mid \mathcal{F}_s\} &= \frac{1}{\xi_s^\theta} I\!E\{\xi_t^\theta e^{iu(W_t^* - W_s^*)} \mid \mathcal{F}_s\} \\
&= e^{\theta W_s + \theta^2 s/2} I\!E\{e^{-\theta W_t - \theta^2 t/2} e^{iu(W_t - W_s + \theta(t-s))} \mid \mathcal{F}_s\} \\
&= e^{(-\theta^2/2 + iu\theta)(t-s)} I\!E\{e^{i(u+i\theta)(W_t - W_s)} \mid \mathcal{F}_s\} \\
&= e^{(-\theta^2/2 + iu\theta)(t-s)} e^{-(u+i\theta)^2(t-s)/2} \\
&= e^{-u^2(t-s)/2}.
\end{aligned}
$$

1.4.2 Self-Financing Portfolios

As in Section 1.3.1, a portfolio comprises a_t units of stock and b_t in bonds; we denote by V_t its value at time t:

$$
V_t = a_t X_t + b_t e^{rt}.
$$

The self-financing property (1.28), namely $dV_t = a_t\, dX_t + rb_t e^{rt}\, dt$, implies that the discounted value of the portfolio, $\widetilde{V}_t = e^{-rt} V_t$, is a martingale under the risk-neutral probability $I\!P^*$. This important property of self-financing portfolios is obtained as follows:

$$
\begin{aligned}
d\widetilde{V}_t &= -re^{-rt} V_t\, dt + e^{-rt}\, dV_t \\
&= -re^{-rt}(a_t X_t + b_t e^{rt})\, dt + e^{-rt}(a_t\, dX_t + rb_t e^{rt}\, dt) \\
&= -re^{-rt} a_t X_t\, dt + e^{-rt} a_t\, dX_t \\
&= a_t\, d(e^{-rt} X_t) \\
&= a_t\, d\widetilde{X}_t \\
&= \sigma a_t \widetilde{X}_t\, dW_t^* \quad \text{(by (1.46)),} \tag{1.50}
\end{aligned}
$$

which shows that $(\widetilde{V}_t)$ is a martingale under $I\!P^*$ as a stochastic integral with respect to the Brownian motion (W_t^*). Indeed, the same computation shows that if a portfolio satisfies $d\widetilde{V}_t = a_t\, d\widetilde{X}_t$ then it is self-financing.

A simple calculation demonstrates the connection between martingales and no arbitrage. Suppose that $(a_t, b_t)_{0 \leq t \leq T}$ is a self-financing arbitrage strategy; that is,

$$V_T \geq e^{rT} V_0 \quad (I\!\!P\text{-a.s.}), \tag{1.51}$$

with

$$I\!\!P\{V_T > e^{rT} V_0\} > 0, \tag{1.52}$$

so that the strategy never makes less than money in the bank and there is some chance of making more. But

$$I\!\!E^\star\{V_T\} = e^{rT} V_0$$

by the martingale property, so (1.51) and (1.52) cannot hold. This is because $I\!\!P$ and $I\!\!P^\star$ are equivalent and so (1.51) and (1.52) also hold with $I\!\!P$ replaced by $I\!\!P^\star$.

1.4.3 Risk-Neutral Valuation

Assume that (a_t, b_t) is a self-financing portfolio satisfying the same integrability conditions of Section 1.3.1 and replicating the European-style derivative with nonnegative payoff H:

$$a_T X_T + b_T e^{rT} = H, \tag{1.53}$$

where we assume that H is a square integrable $\mathcal{F}_T$-adapted random variable. This includes European calls and puts or more general standard European derivatives for which $H = h(X_T)$, as well as other European-style exotic derivatives presented in Section 1.2.3.

On one hand, a no-arbitrage argument shows that the price at time t of this derivative should be the value V_t of this portfolio. On the other hand, as shown in Section 1.4.2, the discounted values $(\widetilde{V}_t)$ of this portfolio form a martingale under the risk-neutral probability $I\!\!P^\star$ and consequently

$$\widetilde{V}_t = I\!\!E^\star\{\widetilde{V}_T \mid \mathcal{F}_t\},$$

which gives

$$V_t = I\!\!E^\star\{e^{-r(T-t)} H \mid \mathcal{F}_t\} \tag{1.54}$$

after reintroducing the discounting factor and using the replicating property (1.53).

Alternatively, given the risk-neutral valuation formula (1.54), we can find a self-financing replicating portfolio for the payoff H. The existence of such a portfolio is guaranteed by an application of the **martingale representation theorem**: for $0 \leq t \leq T$,

$$M_t = I\!\!E^\star\{e^{-rT} H \mid \mathcal{F}_t\}$$

defines a square integrable martingale under $I\!P^\star$ with respect to the filtration $(\mathcal{F}_t)$, which is also the natural filtration of the Brownian motion $W^\star$. This representation theorem says that any such martingale is a stochastic integral with respect to $W^\star$, so that

$$I\!E^\star\{e^{-rT}H \mid \mathcal{F}_t\} = M_0 + \int_0^t \eta_s \, dW_s^\star,$$

where (η_t) is some adapted process with $I\!E^\star\left\{\int_0^T (\eta_t)^2 \, dt\right\}$ finite. By defining $a_t = \eta_t/(\sigma\widetilde{X}_t)$ and $b_t = M_t - a_t\widetilde{X}_t$ we construct a portfolio (a_t, b_t), which is shown to be self-financing by checking that its discounted value is the martingale M_t and using the characterization (1.50) obtained in Section 1.4.2. Its value at time T is $e^{rT}M_T = H$ and hence it is a replicating portfolio.

1.4.4 Using the Markov Property

If H is a function of the path of the stock price after time t – as, for instance, for a standard European derivative with payoff $H = h(X_T)$ – then the Markov property of (X_t) says that conditioning with respect to the past $\mathcal{F}_t$ is the same as conditioning with respect to X_t, the value at the current time; this gives

$$V_t = I\!E^\star\{e^{-r(T-t)}h(X_T) \mid X_t\}.$$

We will come back to this property in the next section.

Denoting by $P(t, x)$ the price of this derivative at time t for an observed stock price $X_t = x$, we obtain the **pricing formula**

$$P(t, x) = I\!E^\star\{e^{-r(T-t)}h(X_T) \mid X_t = x\}. \tag{1.55}$$

If we compare this formula (at time $t = 0$) with (1.23), our first intuitive idea for pricing a standard European derivative in Section 1.2.1, we see that the essential step is to replace the "objective world" $I\!P$ by the "risk-neutral world" $I\!P^\star$ in order to obtain the fair no-arbitrage price.

Knowing that $X_t = x$, one can generalize the formula (1.20) obtained in Section 1.1.5 and obtain an explicit formula for X_T by solving the stochastic differential equation (1.2) from t to T starting from x:

$$X_T = x \exp((\mu - \tfrac{1}{2}\sigma^2)(T - t) + \sigma(W_T - W_t)). \tag{1.56}$$

Using (1.45), this formula can be rewritten in terms of $(W_t^\star)$ as

$$X_T = x \exp((r - \tfrac{1}{2}\sigma^2)(T - t) + \sigma(W_T^\star - W_t^\star)).$$

As $(W_t^\star)$ is a standard Brownian motion under the risk-neutral probability $I\!P^\star$, the increment $W_T^\star - W_t^\star$ is $\mathcal{N}(0, T - t)$-distributed and (1.55) gives the Gaussian integral

$$P(t, x) = \frac{1}{\sqrt{2\pi(T-t)}} \int_{-\infty}^{+\infty} e^{-r(T-t)} h(xe^{(r-\sigma^2/2)(T-t)+\sigma z}) e^{-z^2/2(T-t)} \, dz.$$

$$(1.57)$$

In the case of a European call option, $h(x) = (x - K)^+$, this integral reduces to the Black–Scholes formula (1.39) given in Section 1.3.4.

The two approaches developed in Sections 1.3 and 1.4 should give the same fair price to the same derivative. This is indeed the case and is the content of the following section, where we explain that a formula like (1.55) is just a probabilistic representation of the solution of a partial differential equation like (1.35).

1.5   Risk-Neutral Expectations and Partial Differential Equations

In Section 1.4.4 we used the Markov property of the stock price (X_t) and, in order to compute X_T knowing that $X_t = x$ at time $t \leq T$, we solved the stochastic differential equation (1.2) between t and T. This was a particular case of the general situation where (X_t) is the unique solution of the stochastic differential equation (1.11). We denote by $(X_s^{t,x})_{s \geq t}$ the solution of that equation, starting from x at time t:

$$X_s^{t,x} = x + \int_t^s \mu(u, X_u^{t,x}) \, du + \int_t^s \sigma(u, X_u^{t,x}) \, dW_u,$$

and we assume enough regularity in the coefficients μ and σ for $(X_s^{t,x})$ to be jointly continuous in the three variables (t, x, s). The *flow property* for deterministic differential equations can be extended to stochastic differential equations like (1.11); it says that, in order to compute the solution at time $s > t$ starting at time 0 from point x, one can use

$$x \longrightarrow X_t^{0,x} \longrightarrow X_s^{t, X_t^{0,x}} = X_s^{0,x} \quad (\mathbb{P}\text{-a.s.}). \tag{1.58}$$

In other words, one can solve the equation from 0 to t, starting from x, to obtain $X_t^{0,x}$. Then we solve the equation from t to s, starting from $X_t^{0,x}$. This is the same as solving the equation from 0 to s, starting from x.

The **Markov property** is a consequence and can be stated as follows:

$$\mathbb{E}\{h(X_s) \mid \mathcal{F}_t\} = \mathbb{E}\{h(X_s^{t,x})\}\big|_{x=X_t}, \tag{1.59}$$

which is what we have used with $s = T$ to derive (1.55). Observe that the discounting factor could be pulled out of the conditional expectation since the interest rate is constant (not random). In the time-homogeneous case (μ and σ independent of time) we further have

$$\mathbb{E}\{h(X_s^{t,x})\} = \mathbb{E}\{h(X_{s-t}^{0,x})\},$$

which could have been used with $s = T$ to derive (1.57) since W_{T-t}^* is $\mathcal{N}(0, T - t)$-distributed.

1.5.1   Infinitesimal Generators and Associated Martingales

For simplicity we first consider a time-homogeneous diffusion process (X_t) that solves the stochastic differential equation

$$dX_t = \mu(X_t)\, dt + \sigma(X_t)\, dW_t. \qquad (1.60)$$

Let g be a twice continuously differentiable function of the variable x with bounded derivatives, and define the differential operator $\mathcal{L}$ acting on g according to

$$\mathcal{L}g(x) = \tfrac{1}{2}\sigma^2(x)g''(x) + \mu(x)g'(x). \qquad (1.61)$$

In terms of $\mathcal{L}$, Itô's formula (1.16) gives

$$dg(X_t) = \mathcal{L}g(X_t)\, dt + g'(X_t)\sigma(X_t)\, dW_t,$$

which shows that

$$M_t = g(X_t) - \int_0^t \mathcal{L}g(X_s)\, ds \qquad (1.62)$$

defines a martingale. Consequently, if $X_0 = x$, we obtain

$$\mathbb{E}\{g(X_t)\} = g(x) + \mathbb{E}\left\{ \int_0^t \mathcal{L}g(X_s)\, ds \right\}.$$

Under the assumptions made on the coefficients μ and σ and on the function g, the Lebesgue dominated convergence theorem is applicable and gives

$$\frac{d}{dt}\mathbb{E}\{g(X_t)\}\Big|_{t=0} = \lim_{t\downarrow 0} \frac{\mathbb{E}\{g(X_t)\} - g(x)}{t}$$

$$= \lim_{t\downarrow 0} \mathbb{E}\left\{ \frac{1}{t}\int_0^t \mathcal{L}g(X_s)\, ds \right\} = \mathcal{L}g(x).$$

The differential operator $\mathcal{L}$ given by (1.61) is called the *infinitesimal generator* of the Markov process (X_t).

Considering now a nonhomogeneous diffusion $(\sigma(t, x), \mu(t, x))$ and functions $g(t, x)$ that depend also on time, (1.62) can be generalized by using the full Itô formula (1.16) to yield the martingale

$$M_t = g(t, X_t) - \int_0^t \left(\frac{\partial g}{\partial t} + \mathcal{L}_s g \right)(s, X_s)\, ds, \qquad (1.63)$$

where the infinitesimal generator $\mathcal{L}_t$ is defined by

$$\mathcal{L}_t = \frac{1}{2}\sigma^2(t, x)\frac{\partial^2}{\partial x^2} + \mu(t, x)\frac{\partial}{\partial x} \tag{1.64}$$

and g is any smooth and bounded function. Finally, it is possible to incorporate a discounting factor by using the integration-by-parts formula to compute the differential of $e^{-rt}g(t, X_t)$ and obtain the martingales

$$M_t = e^{-rt}g(t, X_t) - \int_0^t e^{-rs}\left(\frac{\partial g}{\partial t} + \mathcal{L}_s g - rg\right)(s, X_s)\,ds, \tag{1.65}$$

which introduces the *potential* term $-rg$. This can also be generalized to the case of a potential that depends on t and x, with e^{-rt} being replaced by

$$\exp\left(-\int_0^t r(s, X_s)\,ds\right).$$

1.5.2   Conditional Expectations and Parabolic Partial Differential Equations

Suppose that $u(t, x)$ is a solution of the partial differential equation

$$\frac{\partial u}{\partial t} + \frac{1}{2}\sigma^2(t, x)\frac{\partial^2 u}{\partial x^2} + \mu(t, x)\frac{\partial u}{\partial x} - ru = 0 \tag{1.66}$$

with the final condition $u(T, x) = h(x)$, and assume that it is regular enough to apply Itô's formula (1.16). Using (1.65) we deduce that $M_t = e^{-rt}u(t, X_t)$ is a martingale when $\mathcal{L}_t$, given by (1.64), is the infinitesimal generator of the process (X_t) – in other words, when μ and σ^2 are the drift and diffusion coefficients of (X_t).

The martingale property for times t and T reads $I\!E\{M_T \mid \mathcal{F}_t\} = M_t$, which can be rewritten as

$$u(t, X_t) = I\!E\{e^{-r(T-t)}h(X_T) \mid \mathcal{F}_t\},$$

since $u(T, X_T) = h(X_T)$ according to the final condition. Using the Markov property (1.59), we deduce the following probabilistic representation of the solution u:

$$u(t, x) = I\!E\{e^{-r(T-t)}h(X_T^{t,x})\}, \tag{1.67}$$

which may also be written as

$$u(t, x) = I\!E\{e^{-r(T-t)}h(X_T) \mid X_t = x\} \quad \text{or} \quad u(t, x) = I\!E_{t,x}\{e^{-r(T-t)}h(X_T)\}.$$

If r depends on t and x, the discounting factor becomes $\exp(-\int_t^T r(s, X_s)\,ds)$. The representation (1.67) is then called the *Feynman–Kac formula*.

1.5.3   Application to the Black–Scholes Partial Differential Equation

In the previous section we assumed the existence, uniqueness, and regularity of the solution of the partial differential equation (1.66) in order to apply Itô's formula. A sufficient condition for this is that the coefficients μ and σ are regular enough and that the operator $\mathcal{L}_t$ is *uniformly elliptic,* meaning (in this one-dimensional situation) that there exists a positive constant A such that

$$\sigma^2(t, x) \geq A > 0 \quad \text{for every } t \geq 0 \text{ and } x \in \mathcal{D}, \tag{1.68}$$

so that the diffusion coefficient $\sigma^2(t, x)$ cannot become too small. Here $\mathcal{D}$ is the domain of the process (X_t), which may be natural (e.g., $\mathcal{D} = \{x > 0\}$ for the geometric Brownian motion) or imposed externally from other modeling considerations.

When $\mu(t, x) = rx$ and $\sigma(t, x) = \sigma x$ in (1.66), we have the Black–Scholes partial differential equation (1.35) on the domain $\{x > 0\}$. The ellipticity condition (1.68) is clearly not satisfied, since the diffusion coefficient $\sigma^2 x^2$ goes to zero as the state variable approaches zero. We get around this difficulty here (and also in more general situations) with the change of variable $P(t, x) = u(t, y = \log x)$, so that equation (1.35) becomes

$$\frac{\partial u}{\partial t} + \frac{1}{2}\sigma^2 \frac{\partial^2 u}{\partial y^2} + \left(r - \frac{1}{2}\sigma^2\right)\frac{\partial u}{\partial y} - ru = 0, \tag{1.69}$$

to be solved for $0 \leq t \leq T$, $y \in \mathbb{R}$, and with the final condition $u(T, y) = h(e^y)$. The operator

$$\mathcal{L} = \frac{1}{2}\sigma^2 \frac{\partial^2}{\partial y^2} + \left(r - \frac{1}{2}\sigma^2\right)\frac{\partial}{\partial y}$$

is the infinitesimal generator of the (nonstandard) Brownian motion

$$Y_t = (r - \tfrac{1}{2}\sigma^2)t + \sigma W_t^{\star},$$

where $(W_t^{\star})$ is a standard Brownian motion under $I\!P^{\star}$. We use here the same notation as in the equivalent martingale measure context, but the only important fact is that $W^{\star}$ is a standard Brownian motion with respect to the probability used to compute the expectation in the Feynman–Kac formula (1.67). Applying this formula to Y_t yields

$$u(t, y) = I\!E^{\star}\left\{e^{-r(T-t)}h(e^{y+(r-\sigma^2/2)(T-t)+\sigma(W_T^{\star}-W_t^{\star})}) \mid Y_t = y\right\},$$

which is indeed the same as (1.57) by undoing the change of variable $e^y = x$.

1.5.4 American Options and Free Boundary Problems

Pricing American derivatives is mathematically more involved than the European case. Using the theory of *optimal stopping,* it can be shown that the price of an American derivative with payoff function h is obtained by maximizing (over all the stopping times) the expected value of the discounted payoff. As in the European case, the expectations must be taken with respect to the risk-neutral probability in order to avoid arbitrage opportunities. In other words, the intuitive idea presented in Section 1.2.2 is correct when $I\!E$ is replaced by $I\!E^*$:

$$P(0, x) = \sup_{\tau \le T} I\!E^*\{e^{-r\tau}h(X_\tau)\}$$

is the price of the derivative at time $t = 0$, when $X_0 = x$ and where the supremum is taken over all the possible stopping times less than the expiration date T. This formula can be generalized to obtain the price of American derivatives at any time t before expiration T:

$$P(t, x) = \sup_{t \le \tau \le T} I\!E^*\{e^{-r(\tau - t)}h(X_\tau^{t,x})\}, \tag{1.70}$$

where $(X_s^{t,x})_{s \ge t}$ is, as in Section 1.5, the stock price starting at time t from the observed price x.

By taking $\tau = t$ we deduce that $P(t, x) \ge h(x)$, which is natural since if $P(t, x) < h(x)$ then there would be an obvious instant arbitrage at time t. Moreover, by choosing $t = T$ we obtain $P(T, x) = h(x)$.

Because an American derivative gives its holder more rights than the corresponding European derivative, the price of the American is always greater than or equal to the price of the European derivative that has the same payoff function and expiration date. By taking $\tau = T$, we deduce that this property is preserved for $P(t, x)$ defined by (1.70).

Formula (1.70) gives the price of an American derivative. The supremum in (1.70) is reached at the *optimal stopping time,*

$$\tau^* = \tau^*(t) = \inf_s\{t \le s \le T, \ P(s, X_s) = h(X_s)\}, \tag{1.71}$$

the first time that the price of the derivative drops down to its payoff. In order to determine τ^*, one must first compute the price. In terms of partial differential equations, this leads to a *free boundary value problem.* To illustrate, we consider the case of an American put option defined in Section 1.2.2.

It can be shown by a no-arbitrage argument that, for nonnegative interest rates and no dividend paid, the price of an American call option is the same as its corresponding European option. The price of an American put option,

$$P^a(t, x) = \sup_{t \leq \tau \leq T} I\!\!E^* \{ e^{-r(\tau - t)} (K - X_\tau^{t,x})^+ \},$$

is in general strictly higher than the price of the corresponding European put option, which we obtained in closed form in equation (1.42). In fact, we saw in Figure 1.3 that the Black–Scholes put option pricing function crosses below the payoff "ramp" function $(K - x)^+$ for small enough x. This violates $P(t, x) \geq h(x)$, so the European formula for a put cannot also give the price of the American contract, as is the case for call options. We therefore use a put option as our canonical example of an American security in Chapter 9.

Pricing functions for American derivatives satisfy partial differential *inequalities*. For the nonnegative payoff function h, the price of the corresponding American derivative is the solution of the following *linear complementarity problem*:

$$P \geq h,$$

$$\frac{\partial P}{\partial t} + \frac{1}{2}\sigma^2 x^2 \frac{\partial^2 P}{\partial x^2} + rx \frac{\partial P}{\partial x} - rP \leq 0, \qquad (1.72)$$

$$\left(\frac{\partial P}{\partial t} + \frac{1}{2}\sigma^2 x^2 \frac{\partial^2 P}{\partial x^2} + rx \frac{\partial P}{\partial x} - rP \right)(h - P) = 0,$$

to be solved in $\{(t, x) : 0 \leq t \leq T, x > 0\}$ with the final condition $P(T, x) = h(x)$. The second inequality is linked to the supermartingale property of $e^{-rt}P(t, X_t)$ through (1.65) applied to $g = P$.

To see that the price (1.70) is the solution of the differential inequalities (1.72) with the optimal stopping time given by (1.71), assume that we can apply Itô's formula to the solution P of (1.72). For any stopping time $t \leq \tau \leq T$ we have

$$e^{-r\tau}P(\tau, X_\tau^{t,x}) = e^{-rt}P(t, x) + \int_t^\tau e^{-rs} \left(\frac{\partial}{\partial t} + \mathcal{L} - r \right) P(s, X_s^{t,x}) \, ds$$

$$+ \int_t^\tau e^{-rs} \sigma X_s^{t,x} \frac{\partial P}{\partial x}(s, X_s^{t,x}) \, dW_s^*,$$

where $\mathcal{L}$ is the infinitesimal generator of X. The integrand of the Riemann integral is nonpositive by (1.72) and, since τ is bounded, the expectation of the martingale term is zero by Doob's optional stopping theorem. This leads to

$$I\!\!E^* \{ e^{-r(\tau - t)} P(\tau, X_\tau^{t,x}) \} \leq P(t, x)$$

and, using the first inequality in (1.72),

$$I\!\!E^* \{ e^{-r(\tau - t)} h(X_\tau^{t,x}) \} \leq P(t, x).$$

It is easy to see now that if $\tau = \tau^*(t)$, the optimal stopping time defined in (1.71), then we have equalities throughout. This verifies that if (1.72) has a solution to which Itô's formula can be applied then it is the American derivative price (1.70).

In the case of the American put option there is an increasing function $x^*(t)$ – the free boundary – such that, at time t,

$$P(t, x) = K - x \quad \text{for } x < x^*(t),$$

$$\frac{\partial P}{\partial t} + \frac{1}{2}\sigma^2 x^2 \frac{\partial^2 P}{\partial x^2} + rx \frac{\partial P}{\partial x} - rP = 0 \qquad \text{for } x > x^*(t),$$

(1.73)

with

$$P(T, x) = (K - x)^+,$$

(1.74)

$$x^*(T) = K.$$

(1.75)

In addition, P and $\frac{\partial P}{\partial x}$ are continuous across the boundary $x^*(t)$, so that

$$P(t, x^*(t)) = K - x^*(t),$$

(1.76)

$$\frac{\partial P}{\partial x}(t, x^*(t)) = -1.$$

(1.77)

The exercise boundary $x^*(t)$ separates the *hold* region, where the option is not exercised, from the *exercise* region, where it is; this is illustrated in Figure 1.4. In the corresponding Figure 1.5, we show the trajectory of the stock price and the optimal exercise time τ^*.

As in (1.69), the change of variable $y = \log x$ is convenient for analytical and numerical purposes. Notice that this is a system of equations and boundary conditions for $P(t, x)$ *and* the free boundary $x^*(t)$.

1.5.5 Path-Dependent Derivatives

In order to price path-dependent derivatives, one must compute the expectations of their discounted payoffs with respect to the risk-neutral probability. To illustrate this we give two examples.

A **down-and-out call option** (European style) is an example of a barrier option that has a payoff function given by (1.25), and its no-arbitrage price at time $t = 0$ for a stock price equal to x is

$$P(0, x) = I\!E^*\{e^{-rT}(X_T - K)^+ \mathbf{1}_{\{\inf_{t \leq T} X_t > B\}}\}.$$

This can be computed by solving the Black–Scholes partial differential equation (1.35) with the final condition $P(T, x) = (x - K)^+$, corresponding to a call option, and with the boundary condition $P(t, B) = 0$ at $\{x = B\}$ for $t < T$, which

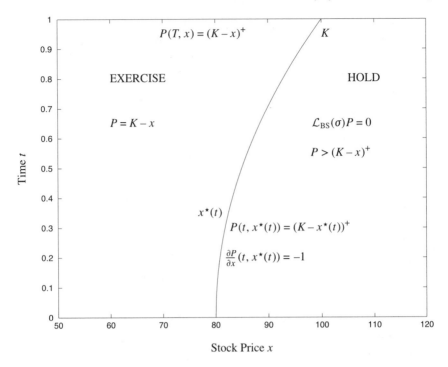

Figure 1.4. The American put problem for $P(t, x)$ and $x^*(t)$, with $\mathcal{L}_{BS}(\sigma)$ defined in (1.36).

ensures that the option is knocked out if the stock price ever falls to $X = B$. Using the *method of images,* one can deduce the value of this barrier option in terms of the Black–Scholes price of the corresponding call option:

$$P(t, x) = C_{BS}(t, x) - (x/B)^{1-2r/\sigma^2} C_{BS}(t, B^2/x).$$

Our second example is an **Asian average-strike option** (European style) whose payoff is given by a function of the stock price at maturity and of the arithmetically averaged stock price before maturity, as in equation (1.26), for example. Without entering into further details, one can introduce the integral process

$$I_t = \int_0^t X_s \, ds$$

and redo the replicating strategies analysis or the risk-neutral valuation argument for the pair of processes (X_t, I_t). Observe that (I_t) does not introduce new risk; in other words, there is no new Brownian motion in the equation $dI_t = X_t \, dt$. Using

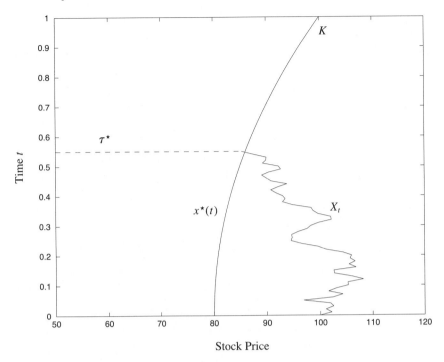

Figure 1.5. Optimal exercise time $\tau^\star$ for an American put option.

a two-dimensional version of Itô's formula, one can deduce the partial differential equation

$$\frac{\partial P}{\partial t} + \frac{1}{2}\sigma^2 x^2 \frac{\partial^2 P}{\partial x^2} + r\left(x\frac{\partial P}{\partial x} - P\right) + x\frac{\partial P}{\partial I} = 0,$$

to be solved with the final condition $P(T, x, I) = (x - I/T)^+$. This is solved numerically in most examples.

1.6   Complete Market

The model we have analyzed here is an example of a *complete* market model. The simplest definition of a complete market is one in which every contingent claim can be replicated by a self-financing trading strategy in the stock and bond.

In Section 1.3.2, we constructed such a strategy for any European derivative with payoff $h(X_T)$, using the Markov property and Itô's lemma, and outlined the arguments for American and Asian contracts. In fact, in this model, *any* security whose payoff H is known on date T (where H is any $\mathcal{F}_T$-measurable random variable with $I\!\!E\{H^2\} < \infty$) can be replicated by some unique self-financing trading strategy, as the martingale representation theorem of Section 1.4.3 tells us. Equivalently, such a claim can be *perfectly* hedged (without overshooting) by trading in the underlying stock and bond. If there is no early exercise feature to the contract (nothing is paid out except on date T and perhaps dividends on *fixed* dates, or at a fixed rate), then the risk-neutral valuation formula (1.54) prices the derivative.

Finally, we mention that another characterization of an arbitrage-free complete market is that there is a *unique* equivalent martingale measure $I\!\!P^*$ under which the discounted prices of traded securities are martingales. When looking at stochastic volatility market models in the next chapter, we shall see that the market is *incomplete*: there is a whole family of equivalent martingale measures, and derivatives securities cannot be perfectly hedged with just the stock and bond.

Notes

The original derivation of the no-arbitrage price of a European call option under the lognormal model appeared in Black and Scholes (1973). The geometric Brownian motion model for the risky asset and many other issues regarding the pricing of options prior to the Black–Scholes theory are discussed in Samuelson (1973).

A good reference for further details about the material outlined in Section 1.1 – namely, Brownian motion, stochastic integrals, stochastic differential equations, and Itô's formulas – is Oksendal (1998). This also has the essential Girsanov's and martingale representation theorems discussed in Sections 1.4.1 and 1.4.3 as well as the optimal stopping theory introduced in Section 1.5.4 for the American option pricing problem, together with an extensive list of further references on the subject.

There are now many books discussing the complete market pricing theory and covering the topics we have summarized in this chapter. Among them are Duffie (1996), Hull (1999), Lamberton and Lapeyre (1996), and Musiela and Rutkowski (1997).

A reference for the method of images approach to pricing barrier options mentioned in Section 1.5.5 is Wilmott, Howison, and Dewynne (1996), which also includes details about the linear complementarity and partial differential inequality formulations of the American pricing problem.

2 Introduction to Stochastic Volatility Models

The Black–Scholes model rests upon a number of assumptions that are, to some extent, "counterfactual." Among these are continuity of the stock-price process (it does not jump), the ability to hedge continuously without transaction costs, independent Gaussian returns, and constant volatility. We shall focus here on relaxing the last assumption by allowing volatility to vary randomly, for the following reason: a well-known discrepancy between Black–Scholes-predicted European option prices and market-traded options prices, the **smile curve**, can be accounted for by stochastic volatility models. That is, this modification of the Black–Scholes theory has a posteriori success in one area where the classical model fails.

In fact, modeling volatility as a stochastic process is motivated a priori by empirical studies of stock-price returns in which estimated volatility is observed to exhibit "random" characteristics. Additionally, the effects of transaction costs show up, under many models, as uncertainty in the volatility; fat-tailed returns distributions can be simulated by stochastic volatility; and market "jump" phenomena are often best modeled as volatility jump processes. Stochastic volatility modeling therefore is not just a simple fix to one particular Black–Scholes assumption but rather a powerful modification that describes a much more complex market. We cite literature that explores possible causes of stochastic volatility in the notes at the end of this chapter.

In Chapter 1, we introduced the notation and tools for pricing and hedging derivative securities under a constant volatility lognormal model (1.2). This is the simplest example of pricing in a complete market. However, pricing in a market with stochastic volatility is an incomplete markets problem, a distinction that (as we shall explain) has far-reaching consequences – particularly for the hedging problem and the problem of parameter estimation. It is the latter *inverse* problem that is the biggest mathematical and practical challenge introduced by such models, and also perhaps the one that benefits most from the asymptotic methods of Chapter 5.

2.1   Implied Volatility and the Smile Curve

Implied volatility is a convenient synoptic variable frequently used to express differences between Black–Scholes European options prices and market options prices.

Given an observed European call option price C^{obs} for a contract with strike price K and expiration date T, the *implied volatility* I is defined to be the value of the volatility parameter that must go into the Black–Scholes formula (1.37) to match this price:

$$C_{BS}(t, x; K, T; I) = C^{obs}. \tag{2.1}$$

Remarks. (1) A unique nonnegative implied volatility $I > 0$ can be found given $C^{obs} > C_{BS}(t, x; K, T; 0)$ because of the monotonicity of the Black–Scholes formula in the volatility parameter:

$$\frac{\partial C_{BS}}{\partial \sigma} = \frac{xe^{-d_1^2/2}\sqrt{T - t}}{\sqrt{2\pi}} > 0. \tag{2.2}$$

(2) The implied volatilities from put and call options of the same strike price and time to maturity are the same because of put–call parity (1.41).

At first glance, implied volatility seems a symptom of wishful thinking that answers the question: How can the input volatility parameter be tweaked to make the Black–Scholes formula work? It turns out, however, to be a useful quantity with which to compare model predictions and observations. Many formulas, for example the asymptotic ones of Chapter 5, are most neatly expressed in terms of implied volatility. Traders also often quote derivative prices in terms of I, the conversion to price being through the Black–Scholes formula appropriate for the contract.

In general, $I = I(t, x; K, T)$, but if the observed options prices equaled the Black–Scholes prices, $C^{obs} = C_{BS}(t, x; K, T; \sigma)$, then $I = \sigma$, the historical volatility. That is, if observations matched Black–Scholes exactly, I would be the same constant for all derivative contracts.

The most quoted phenomenon testifying to the limitations of the standard Black–Scholes model is the *smile effect*: that implied volatilities of market prices are not constant across the range of options but vary with strike price and the time to maturity of the contract. Before the 1987 crash, the graph of $I(K)$ against K for fixed t, x, T obtained from market options prices was often observed to be U-shaped, with minimum at or near the money ($K_{min} \approx x$). This is called the smile curve, illustrated in Figure 2.1. Since that time, the curve is more typically downward sloping at and near the money ($95\% \leq K/x \leq 105\%$) and then curves upwards for far out-of-the-money strikes ($K \gg x$). An example from S&P 500 index options is shown in Figure 2.2.

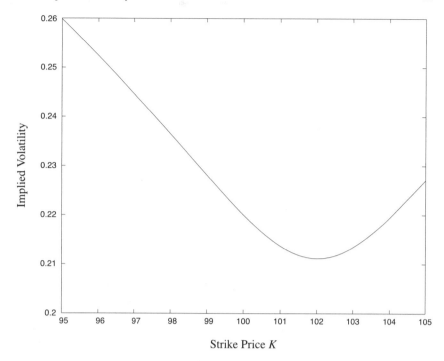

Figure 2.1. Illustrative smile curve of implied volatilities from European options with the same time to expiration. The current stock price is $x = 100$, which is close to the minimum point. In fact, the minimum is at $xe^{r(T-t)} \approx 102$.

Other qualitative features of implied volatility from stock options are that it is higher than historical volatility and is often decreasing with time to maturity.

2.1.1   Interpretation of the Smile Curve

Such a smile curve, or **smirk**, tells us that there is a premium charged for out-of-the-money put options and in-the-money calls (low K) above their Black–Scholes price computed with at-the-money implied volatility. The market itself prices as though the lognormal model fails to capture probabilities of large downward stock price movements and so supplements the Black–Scholes prices to account for this.

One can obtain some broad bounds on the permissible slope of the implied volatility curve $I(K)$ by noting that call prices must be decreasing in the strike price K (or else there is an arbitrage opportunity). Differentiating (2.1) with respect to K gives

$$\frac{\partial C^{\text{obs}}}{\partial K} = \frac{\partial C_{\text{BS}}}{\partial K} + \frac{\partial C_{\text{BS}}}{\partial \sigma}\frac{\partial I}{\partial K} \leq 0,$$

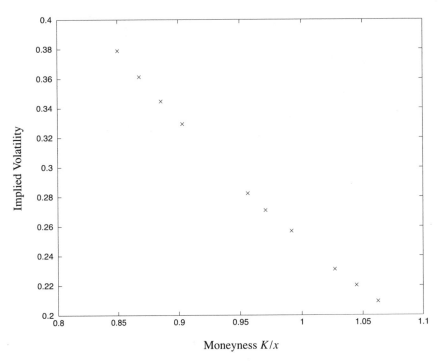

Figure 2.2. S&P 500 implied volatility curve as a function of moneyness from S&P 500 index options on February 9, 2000. The current index value is $x = 1411.71$ and the options have over two months to maturity. This is typically described as a downward sloping *skew*.

from which we conclude that

$$\frac{\partial I}{\partial K} \leq -\frac{\partial C_{BS}/\partial K}{\partial C_{BS}/\partial \sigma}.$$

Similarly, put prices must be increasing in K. Because prices of puts and calls with the same K and T must have the same implied volatility I, we also have

$$\frac{\partial I}{\partial K} \geq -\frac{\partial P_{BS}/\partial K}{\partial P_{BS}/\partial \sigma}.$$

Using the explicit Black–Scholes formulas (1.37), we obtain the bounds

$$-\frac{\sqrt{2\pi}}{x\sqrt{T-t}}(1 - N(d_2))e^{-r(T-t)+d_1^2/2} \leq \frac{\partial I}{\partial K} \leq \frac{\sqrt{2\pi}}{x\sqrt{T-t}}N(d_2)e^{-r(T-t)+d_1^2/2},$$

where d_1 and d_2 are defined in (1.38)–(1.39) with σ replaced by I. In other words, the slope cannot be too positive or too negative.

This type of calculation, where information about the geometry of the smile curve is obtained from the Black–Scholes formula and its derivatives, will come up often – for example, to show that stochastic volatility models predict the smile (Section 2.8.2) and in the asymptotic calculations of Chapter 5.

At this stage, it is clear that the specification of a constant volatility is significantly inaccurate, first from statistical studies of stock price history that indicate the random character of volatility (measured as the normalized standard deviation of returns and discussed in Chapter 4) and second from the nonflat implied volatility surface $I(K, T)$, which tells us how the market is actually pricing derivatives contrary to the Black–Scholes theory.

2.1.2 What Data to Use

When it comes to estimation, there is a radical departure from the spirit of the Black–Scholes model both in practice and in the literature.

It has long been common practice in the industry to use implied volatility as a proxy for volatility. In other words, options are priced with the Black–Scholes formula but using implied volatilities instead of historical volatilities. In particular, out-of-the-money puts ($K < x$) are typically priced with a higher volatility that may be derived from the previous day's smile curve. The smile curve is thus translated forward in time, and no historical stock-price data other than the current price is used at all. Today's options are priced using yesterday's *options* data.

The main reason for this procedure – which is, of course, mathematically inconsistent since the Black–Scholes formula assumes constant volatility (independent of the derivative contract) in its derivation – is the belief that implied volatilities are better predictors of future realized volatility. Empirical studies disagree as to whether this is borne out by the data.

In tests of new models in the academic literature, calibration of parameters is also usually from derivative data (traded call option prices, for example), ignoring the underlying time series. This is because such *cross-sectional fitting* is easier than employing econometric methods for time series, especially when there is a formula to which call option prices can be fitted. Once the parameters have been estimated, the model can be used to price *other* derivatives in a consistent (no-arbitrage) manner – for example, by simulation.

Thus some derivative data, most often at-the-money European option prices, are part of the basic data. The market in at-the-money calls (and puts) is liquid enough for this to be a valid procedure (i.e., we can trust the data).

When we look at stochastic volatility models, we shall see that derivative data contains some information that we need for pricing that is not contained in historical data, so we shall always have to use it in part. We address this further in Section 2.7.

2.2   Implied Deterministic Volatility

One popular way to modify the lognormal model is to suppose that volatility is a deterministic positive function of time and stock price: $\sigma = \sigma(t, X_t)$. The stochastic differential equation modeling the stock price is

$$dX_t = \mu X_t \, dt + \sigma(t, X_t) X_t \, dW_t,$$

and the function $P(t, x)$, giving the no-arbitrage price of a European derivative security at time t when the asset price $X_t = x$, then satisfies the generalized Black–Scholes partial differential equation

$$\frac{\partial P}{\partial t} + \frac{1}{2}\sigma^2(t, x)x^2\frac{\partial^2 P}{\partial x^2} + r\left(x\frac{\partial P}{\partial x} - P\right) = 0,$$

the derivation being identical to that given in Section 1.3.2 for the constant-σ case. The coefficient σ becomes $\sigma(t, X_t)$ in equations (1.31) and (1.34), and $\sigma(t, x)$ in (1.35). The terminal condition is the payoff function: $P(T, x) = h(x)$.

The hedging ratio is given by the Delta of the solution to this partial differential equation problem, $\partial P/\partial x$, and a perfect hedge is achieved by holding this amount of stock. This can be seen by repeating the argument of Section 1.3.2, replacing the constant σ coefficient by $\sigma(t, X_t)$.

The market is still complete as the randomness of the volatility was introduced as a function of the existing randomness of the lognormal model. There is a unique risk-neutral measure $I\!P^\star$ under which the stock price is a geometric Brownian motion with drift rate r and the same volatility $\sigma(t, X_t)$:

$$dX_t = rX_t \, dt + \sigma(t, X_t) X_t \, dW_t^\star,$$

with $(W_t^\star)$ a $I\!P^\star$-Brownian motion. Note, however, that the market is *not* complete if the volatility is modeled to have a random component of its own, as in the stochastic volatility models of Section 2.3.

2.2.1   *Time-Dependent Volatility*

In the special case $\sigma(t, x) = \sigma(t)$, a deterministic function of time, we can solve the stochastic differential equation

$$dX_t = rX_t \, dt + \sigma(t) X_t \, dW_t^\star,$$

analogously to the calculation in Section 1.1.5, by the logarithmic transformation to obtain

$$X_T = X_t \exp\left(r(T - t) - \frac{1}{2}\int_t^T \sigma^2(s)\, ds + \int_t^T \sigma(s)\, dW_s^\star\right),$$

so that $\log(X_T/X_t)$ is $\mathcal{N}\big((r - \frac{1}{2}\overline{\sigma^2})(T - t), \overline{\sigma^2}(T - t)\big)$-distributed, where

$$\overline{\sigma^2} = \frac{1}{T - t} \int_t^T \sigma^2(s)\, ds.$$

When we compute the call option price from

$$C = I\!\!E^\star\{e^{-r(T-t)}(X_T - K)^+ \mid \mathcal{F}_t\},$$

the answer will just be the Black–Scholes formula with volatility parameter $\sqrt{\overline{\sigma^2}}$, the root-mean-square (RMS) volatility.

For fixed t and T, all options are Black–Scholes priced with the same (time-averaged) volatility, so there is no smile across strike prices. There is, however, variation of implied volatility with time to maturity since $\overline{\sigma^2}$ is different for different maturities. This feature of pricing using Black–Scholes with an averaged volatility (in particular, the square root of the average of squared volatility) is something we shall see many times when we come to asymptotic approximations of stochastic volatility prices in Chapter 5.

2.2.2   Level-Dependent Volatility

To have a smile across strike prices, we need σ to depend on x as well as t in this framework: $\sigma(t, X_t)$. One disadvantage is that volatility and stock price changes are now perfectly correlated (positively or negatively). Empirical studies suggest that prices tend to go down when volatility goes up and vice versa, but there is not a perfect (-1) inverse correlation.

2.2.3   Short-Time Tight Fit versus Long-Time Rough Fit

There are many competing ways – some parametric and some nonparametric – to estimate the volatility surface $\sigma(t, x)$ from traded option prices. This is called "finding the implied deterministic volatility." It has the advantage of preserving a complete market model. Where the problem lies is in the stability of the fits over time: with new data a week later, the fits are often completely different even though, given the large degree of freedom, they are very good fits to the current data. References for empirical evidence are given at the end of the chapter.

There is a trade-off between a tight fit over a short time versus a rough fit over a longer time, and the problem of parameter stability is an important criterion with which to assess any model.

2.3   Stochastic Volatility Models

In "pure" stochastic volatility models (as opposed to implied deterministic volatility), the asset price $(X_t)_{t\geq 0}$ satisfies the stochastic differential equation

$$dX_t = \mu X_t \, dt + \sigma_t X_t \, dW_t, \tag{2.3}$$

where $(\sigma_t)_{t\geq 0}$ is called the volatility process. It must satisfy some regularity conditions for the model to be well-defined, but it does not have to be an Itô process: it can be a jump process, a Markov chain, ... In order for it to be a volatility, it should be positive. Unlike the implied deterministic volatility models, the volatility process is *not* perfectly correlated with the Brownian motion (W_t). Therefore, volatility is modeled to have an independent random component of its own. We shall see in a number of ways that this leads to an incomplete market and that there is no *unique* equivalent martingale measure.

2.3.1   *Mean-Reverting Stochastic Volatility Models*

Typically, volatility is taken to be an Itô process satisfying a stochastic differential equation driven by a second Brownian motion. This is the easiest way to incorporate correlation with stock-price changes. Within this framework, we have finite-difference or tree methods available for computation. After that, we want a model in which volatility is positive; some choices are lognormal or Feller (CIR) processes, described in this section.

One feature that most models seem to like is *mean reversion*. The term "mean reverting" refers to the characteristic (typical) time it takes for a process to get back to the mean level of its invariant distribution (the long-run distribution of the process). This is discussed in more detail in Chapter 3.

From a financial modeling perspective, mean reverting refers to a linear pullback term in the drift of the volatility process itself, or in the drift of some (underlying) process of which volatility is a function. Let us denote $\sigma_t = f(Y_t)$, where f is some positive function. Then mean-reverting stochastic volatility means that the stochastic differential equation for (Y_t) looks like

$$dY_t = \alpha(m - Y_t) \, dt + \cdots d\hat{Z}_t,$$

where $(\hat{Z}_t)_{t\geq 0}$ is a Brownian motion correlated with (W_t). Here α is called the *rate of mean reversion* and m is the long-run mean level of Y. The drift term pulls Y toward m, so we would expect that σ_t is pulled toward the mean value of $f(Y)$ with respect to the long-run distribution of Y.

For example, the **Ornstein–Uhlenbeck process** is defined as a solution of

$$dY_t = \alpha(m - Y_t)\,dt + \beta\,d\hat{Z}_t. \tag{2.4}$$

It is a Gaussian process explicitly given in terms of its (assumed known) starting value y by

$$Y_t = m + (y - m)e^{-\alpha t} + \beta \int_0^t e^{-\alpha(t-s)}\,d\hat{Z}_s, \tag{2.5}$$

so that Y_t is $\mathcal{N}\big(m + (y - m)e^{-\alpha t}, \frac{\beta^2}{2\alpha}(1 - e^{-2\alpha t})\big)$-distributed. Its invariant distribution, obtained as $t \to \infty$, is $\mathcal{N}(m, \beta^2/2\alpha)$, which does not depend on y.

The second Brownian motion $(\hat{Z}_t)$ is typically correlated with the Brownian motion (W_t) driving the asset price equation (2.3). We denote by $\rho \in [-1, 1]$ the instantaneous correlation coefficient defined by

$$d\langle W, \hat{Z} \rangle_t = \rho\,dt,$$

using the notation of (1.19). It is also convenient to write

$$\hat{Z}_t = \rho W_t + \sqrt{1 - \rho^2}\,Z_t, \tag{2.6}$$

where (Z_t) is a standard Brownian motion independent of (W_t). It is often found from financial data that $\rho < 0$, and there are economic arguments for a negative correlation or *leverage effect* between stock price and volatility shocks. From common experience and empirical studies, asset prices tend to go down when volatility goes up. In general, the correlation may depend on time $\rho(t) \in [-1, 1]$, but we shall assume it a constant from now on for notational simplicity and because, in most practical situations, it is taken to be such.

We continue to denote the underlying probability space by $(\Omega, \mathcal{F}, I\!P)$, where now we can take $\Omega = \mathcal{C}([0, \infty) : I\!R^2)$, the space of all continuous trajectories $(W_t(\omega), Z_t(\omega)) = \omega(t)$ in $I\!R^2$. The filtration $(\mathcal{F}_t)_{t\geq 0}$ represents the information on the two Brownian motions; so, for example, $\mathcal{F}_t$ is the σ-algebra generated by sets of the form $\{\omega \in \Omega : |W_s| < R_1, |Z_s| < R_2, s \leq t\}$ completed by the null sets.

Some common driving processes (Y_t) are:

- lognormal (LN),

$$dY_t = c_1 Y_t\,dt + c_2 Y_t\,d\hat{Z}_t;$$

- Ornstein-Uhlenbeck (OU),

$$dY_t = \alpha(m - Y_t)\,dt + \beta\,d\hat{Z}_t;$$

Table 2.1: *Models of Volatility*

Authors	Correlation	$f(y)$	Y Process
Hull–White	$\rho = 0$	$f(y) = \sqrt{y}$	Lognormal
Scott	$\rho = 0$	$f(y) = e^y$	Mean-reverting OU
Stein–Stein	$\rho = 0$	$f(y) = |y|$	Mean-reverting OU
Ball–Roma	$\rho = 0$	$f(y) = \sqrt{y}$	CIR
Heston	$\rho \neq 0$	$f(y) = \sqrt{y}$	CIR

• Feller or Cox–Ingersoll–Ross (CIR),

$$dY_t = \kappa(m' - Y_t)\,dt + v\sqrt{Y_t}\,d\hat{Z}_t.$$

Note that the lognormal model is *not* mean-reverting.

Some models studied in the literature are listed in Table 2.1. These particular models are chosen for their nice properties (for example, positivity and mean reversion) and analytical tractability rather than for deeper financial reasons. From the point of view of this book, we will try to develop a framework for stochastic volatility that does not depend upon specific modeling of the volatility process. Our main requirement for the asymptotic theory developed in this book is that the stochastic volatility is driven by an *ergodic* process (explained in Chapter 3), of which the OU process is a perfect example.

2.3.2   Stock-Price Distribution under Stochastic Volatility

What effect does stochastic volatility have on the probability density function of the stock price? To answer this question qualitatively, we plot in Figure 2.3 the density function (estimated from simulation) of the "expOU" stochastic volatility model in which $f(y) = e^y$ and (Y_t) is a mean-reverting OU process. Notice the fatter tails due to the random volatility. In particular, the negative correlation causes the tails to be asymmetric: the left tail is fatter.

2.4   Derivative Pricing

When the volatility is a Markov Itô process, we can find a pricing function for European derivatives of the form $P(t, X_t, Y_t)$ from no-arbitrage arguments, as in the Black–Scholes case. The function $P(t, x, y)$ satisfies a partial differential equation with *two* space dimensions (x and y); the price of the derivative depends on the value of the process y, which is not directly observable. We return to this issue later.

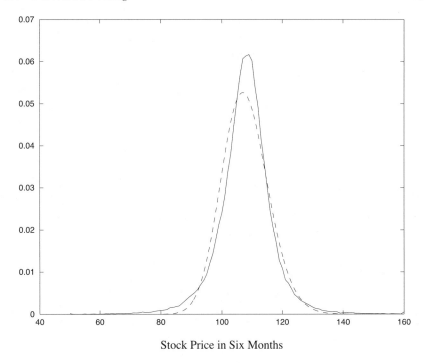

Stock Price in Six Months

Figure 2.3. Density functions for the stock price (under the subjective measure) in six months when the present value is 100. The solid line is estimated from simulation of an expOU stochastic volatility model with $\alpha = 1$, $\beta = \sqrt{2}$, long-run average volatility $\bar{\sigma} = 0.1$, and negative correlation $\rho = -0.2$. The dotted line is the corresponding Black–Scholes lognormal density function with volatility $\bar{\sigma}$. The mean growth rate of the stock is $\mu = 0.15$.

We will derive the pricing partial differential equation assuming that volatility is a function of a mean-reverting OU process:

$$dX_t = \mu X_t \, dt + \sigma_t X_t \, dW_t,$$
$$\sigma_t = f(Y_t), \tag{2.7}$$
$$dY_t = \alpha(m - Y_t) \, dt + \beta \, d\hat{Z}_t,$$

and recall from (2.6) that we can write $(\hat{Z}_t)$ as a linear combination of (W_t) and an independent Brownian motion (Z_t). This keeps the notation simple, but the derivation applies for any Markovian Itô driving process

$$dY_t = \mu_Y(t, Y_t) \, dt + \sigma_Y(t, Y_t) \, d\hat{Z}_t.$$

We then look for the pricing function $P(t, x, y)$ by trying to construct a hedged portfolio of assets that can be priced by the no-arbitrage principle. Unlike the Black–Scholes case, it is not sufficient to hedge solely with the underlying asset, since the dW_t term can be balanced but the dZ_t term cannot. Thus, we try to hedge with the underlying asset and *another* option that has a different expiration date.

Let $P^{(1)}(t, x, y)$ be the price of a European derivative with expiration date T_1 and payoff function $h(X_{T_1})$, and try to find processes $\{a_t, b_t, c_t\}$ such that

$$P^{(1)}(T_1, X_{T_1}, Y_{T_1}) = a_{T_1} X_{T_1} + b_{T_1} \beta_{T_1} + c_{T_1} P^{(2)}(T_1, X_{T_1}, Y_{T_1}), \qquad (2.8)$$

where $\beta_t = e^{rt}$ is the price of a riskless bond under the prevailing short-term constant interest rate r, and $P^{(2)}(t, X_t, Y_t)$ is the price of a European contract with the same payoff function h as $P^{(1)}$ but with a different expiration date $T_2 > T_1 > t$. In other words, the right-hand side of (2.8) is a portfolio whose payoff at time T_1 equals (almost surely) the payoff of $P^{(1)}$. In addition, the portfolio is to be self-financing, so that

$$dP^{(1)}(t, X_t, Y_t) = a_t \, dX_t + b_t r e^{rt} \, dt + c_t \, dP^{(2)}(t, X_t, Y_t). \qquad (2.9)$$

If such a portfolio can be found then, in order for there to be no arbitrage opportunities, it must be that

$$P^{(1)}(t, X_t, Y_t) = a_t X_t + b_t e^{rt} + c_t P^{(2)}(t, X_t, Y_t) \qquad (2.10)$$

for all $t < T_1$.

We shall use the two-dimensional version of Itô's formula (1.15),

$$dg(t, X_t, Y_t) = \frac{\partial g}{\partial t} \, dt + \frac{\partial g}{\partial x} \, dX_t + \frac{\partial g}{\partial y} \, dY_t$$

$$+ \frac{1}{2} \left(\frac{\partial^2 g}{\partial x^2} \, d\langle X \rangle_t + 2 \frac{\partial^2 g}{\partial x \partial y} \, d\langle X, Y \rangle_t + \frac{\partial^2 g}{\partial y^2} \, d\langle Y \rangle_t \right), \qquad (2.11)$$

with the same bracket notation of (1.19).

Applying this to both sides of (2.9) yields

$$\left(\frac{\partial P^{(1)}}{\partial t} + \mathcal{M}_1 P^{(1)} \right) dt + \frac{\partial P^{(1)}}{\partial x} \, dX_t + \frac{\partial P^{(1)}}{\partial y} \, dY_t$$

$$= \left(a_t + c_t \frac{\partial P^{(2)}}{\partial x} \right) dX_t + c_t \frac{\partial P^{(2)}}{\partial y} \, dY_t$$

$$+ \left[c_t \left(\frac{\partial}{\partial t} + \mathcal{M}_1 \right) P^{(2)} + b_t r e^{rt} \right] dt, \qquad (2.12)$$

where

$$\mathcal{M}_1 = \frac{1}{2} f(y)^2 x^2 \frac{\partial^2}{\partial x^2} + \rho \beta x f(y) \frac{\partial^2}{\partial x \partial y} + \frac{1}{2} \beta^2 \frac{\partial^2}{\partial y^2}$$

and where $P^{(1)}$, $P^{(2)}$ and their derivatives are evaluated at (t, X_t, Y_t). Equating the $d\hat{Z}_t$ terms is equivalent to equating the dY_t terms, which gives

$$c_t = \frac{\partial P^{(1)}/\partial y}{\partial P^{(2)}/\partial y}; \tag{2.13}$$

from the dW_t terms associated with dX_t, we must then have

$$a_t = \frac{\partial P^{(1)}}{\partial x} - c_t \frac{\partial P^{(2)}}{\partial x}. \tag{2.14}$$

Substituting for a_t, c_t, and $b_t = (P_t^{(1)} - a_t X_t - c_t P_t^{(2)})/e^{rt}$ and comparing dt terms in (2.12) gives

$$\left(\frac{\partial P^{(1)}}{\partial y} \right)^{-1} \mathcal{M}_2 P^{(1)}(t, X_t, Y_t) = \left(\frac{\partial P^{(2)}}{\partial y} \right)^{-1} \mathcal{M}_2 P^{(2)}(t, X_t, Y_t), \tag{2.15}$$

where

$$\mathcal{M}_2 = \frac{\partial}{\partial t} + \mathcal{M}_1 + r \left(x \frac{\partial}{\partial x} - \cdot \right).$$

That is, $\mathcal{M}_2$ is the standard Black–Scholes differential operator with volatility parameter $f(y)$, plus second-order terms from the Y diffusion process. Now, the left-hand side of (2.15) contains terms depending on T_1 but not T_2 and vice versa for the right-hand side. Thus, both sides must be equal to a function that does not depend on expiration date. We denote this function by

$$\alpha(m - y) - \beta \left(\rho \frac{(\mu - r)}{f(y)} + \gamma(t, x, y) \sqrt{1 - \rho^2} \right),$$

for reasons explained below. Here $\gamma(t, x, y)$ is an arbitrary function.

The pricing function $P(t, x, y)$, with the dependence on expiry date supressed, must satisfy the partial differential equation

$$\frac{\partial P}{\partial t} + \frac{1}{2} f(y)^2 x^2 \frac{\partial^2 P}{\partial x^2} + \rho \beta x f(y) \frac{\partial^2 P}{\partial x \partial y} + \frac{1}{2} \beta^2 \frac{\partial^2 P}{\partial y^2}$$

$$+ r \left(x \frac{\partial P}{\partial x} - P \right) + (\alpha(m - y) - \beta \Lambda(t, x, y)) \frac{\partial P}{\partial y} = 0, \tag{2.16}$$

where

$$\Lambda(t, x, y) = \rho \frac{(\mu - r)}{f(y)} + \gamma(t, x, y) \sqrt{1 - \rho^2}. \tag{2.17}$$

The terminal condition is $P(T, x, y) = h(x)$. Since (Y_t) is an OU process, the y-domain is $(-\infty, \infty)$.

We can group the differential operator on the left side of (2.16) in the following way:

$$
\underbrace{\frac{\partial}{\partial t} + \frac{1}{2}f(y)^2 x^2 \frac{\partial^2}{\partial x^2} + r\left(x\frac{\partial}{\partial x} - \cdot\right)}_{\mathcal{L}_{BS}(f(y))} + \underbrace{\rho\beta x f(y)\frac{\partial^2}{\partial x \partial y}}_{\text{correlation}}
$$

$$
+ \underbrace{\frac{1}{2}\beta^2 \frac{\partial^2}{\partial y^2} + \alpha(m-y)\frac{\partial}{\partial y}}_{\mathcal{L}_{OU}} - \underbrace{\beta\Lambda\frac{\partial}{\partial y}}_{\text{premium}} .
$$

The first grouping is the Black–Scholes operator (1.36) with volatility level $f(y)$; the second is the term due to the correlation; the third is the infinitesimal generator of the OU process (Y_t), following the definition (1.61). The final term is due to the market price of volatility risk.

The function γ in (2.17) is the risk premium factor from the *second* source of randomness (Z_t) that drives the volatility: in the perfectly correlated case $|\rho| = 1$ it does not appear, as expected. The reason for this terminology is the calculation

$$
dP(t, X_t, Y_t) = \left[\frac{(\mu - r)}{f(y)}\left(xf(y)\frac{\partial P}{\partial x} + \beta\rho\frac{\partial P}{\partial y}\right) + rP + \gamma\beta\sqrt{1-\rho^2}\frac{\partial P}{\partial y}\right] dt
$$

$$
+ \left(xf(y)\frac{\partial P}{\partial x} + \beta\rho\frac{\partial P}{\partial y}\right) dW_t + \beta\sqrt{1-\rho^2}\frac{\partial P}{\partial y}\, dZ_t,
$$

which is obtained using the Itô formula (2.11) and the partial differential equation (2.16) satisfied by P. From this expression we see that an infinitesimal fractional increase in the volatility risk β increases the infinitesimal rate of return on the option by γ times that fraction, in addition to the increase from the *excess return-to-risk ratio* $(\mu - r)/f(y)$.

2.5   Pricing with Equivalent Martingale Measures

We give an alternative derivation of the no-arbitrage derivative price using the risk-neutral theory, again for the model (2.7), but the procedure is valid for general models, including non-Markovian ones.

Suppose first that there *is* an equivalent martingale measure $I\!P^\star$ under which the discounted stock price $\widetilde{X}_t = e^{-rt}X_t$ is a martingale. Then we know from Chapter 1 that if we price any (and all) derivatives with time-T payoff H using the formula

$$
V_t = I\!E^\star\{e^{-r(T-t)}H \mid \mathcal{F}_t\}
$$

for all $t \le T$, then there is no arbitrage opportunity. Thus V_t is a possible price for the claim. Let us now try to construct equivalent martingale measures.

As in Section 1.4.1, we absorb the drift term of $\widetilde{X}_t$ in its martingale term by setting

$$W_t^{\star} = W_t + \int_0^t \frac{(\mu - r)}{f(Y_s)}\, ds.$$

Any shift of the second independent Brownian motion of the form

$$Z_t^{\star} = Z_t + \int_0^t \gamma_s\, ds$$

will not change the drift of $\widetilde{X}_t$. By Girsanov's theorem, $(W^{\star})$ and $(Z^{\star})$ are independent standard Brownian motions under a measure $I\!P^{\star(\gamma)}$ defined by

$$\frac{dI\!P^{\star(\gamma)}}{dI\!P} = \exp\left(-\frac{1}{2}\int_0^T ((\theta_s^{(1)})^2 + (\theta_s^{(2)})^2)\, ds - \int_0^T \theta_s^{(1)}\, dW_s - \int_0^T \theta_s^{(2)}\, dZ_s\right),$$

$$\theta_t^{(1)} = \frac{(\mu - r)}{f(Y_t)},$$

$$\theta_t^{(2)} = \gamma_t.$$

Here (γ_t) is *any* adapted (and suitably regular) process. Technically, we shall make an assumption on the pair $\left(\frac{(\mu-r)}{f(Y_t)}, \gamma_t\right)$ so that $I\!P^{\star(\gamma)}$ is well-defined as a probability measure. In particular, this will be the case if f is bounded away from zero and (γ_t) is bounded. Then, under $I\!P^{\star(\gamma)}$, the stochastic differential equations (2.7) become

$$dX_t = rX_t\, dt + f(Y_t)X_t\, dW_t^{\star}, \tag{2.18}$$

$$dY_t = \left[\alpha(m - Y_t) - \beta\left(\rho\frac{(\mu - r)}{f(Y_t)} + \gamma_t\sqrt{1 - \rho^2}\right)\right] dt + \beta\, d\hat{Z}_t^{\star}, \tag{2.19}$$

$$\hat{Z}_t^{\star} = \rho W_t^{\star} + \sqrt{1 - \rho^2} Z_t^{\star}.$$

Any allowable choice of γ leads to an equivalent martingale measure $I\!P^{\star(\gamma)}$ and the possible no arbitrage derivative prices

$$V_t = I\!E^{\star(\gamma)}\{e^{-r(T-t)}H \mid \mathcal{F}_t\}. \tag{2.20}$$

The process (γ_t) is called the *risk premium factor* or the **market price of volatility risk** from the second source of randomness Z that drives the volatility. It parameterizes the space of equivalent martingale measures $\{I\!P^{\star(\gamma)}\}$. When $\gamma = \gamma(t, X_t, Y_t)$, a function of the processes, we are back to the Markovian setting; then (2.16) is just the Kolmogorov (Feynman–Kac) partial differential equation for (2.20).

Note also that we cannot replicate the claim by trading in stock and bond only. Recall that – because we are modeling with two Brownian motions – we are dealing with a larger probability space $\Omega = C([0, \infty) : \mathbb{R}^2)$, and the filtration $(\mathcal{F}_t)$ is generated by (W_t, Z_t). As a result, the martingale representation theorem used in Section 1.4.3 says only that the $\mathbb{P}^{\star(\gamma)}$ martingale $M_t = e^{-rt}V_t$ is a stochastic integral with respect to $(W^\star, Z^\star)$:

$$M_t = M_0 + \int_0^t \eta_s^{(1)} \, dW_s^\star + \int_0^t \eta_s^{(2)} \, dZ_s^\star \qquad (2.21)$$

for some adapted (and suitably bounded) processes $(\eta_t^{(1)})$ and $(\eta_t^{(2)})$. We can translate this into a self-financing trading strategy in stock, bond, *and volatility* (by undoing the discounting and substituting for $dW^\star$ and $dZ^\star$ from (2.18) and (2.19)):

$$dV_t = a_t \, dX_t + b_t r e^{rt} \, dt + c_t \, d\sigma_t$$

for some (a_t, b_t, c_t). However, because of the last integral in (2.21), we cannot write this as a strategy in just the stock and bond ($c_t \neq 0$ in general) as we would like, since volatility is not a tradeable asset.

We can, however, hedge one derivative contract $P^{(1)}$ with the stock and another derivative security $P^{(2)}$, as in equation (2.9). The calculation of Section 2.4 (in the Markovian case) yields the hedging ratios (2.13) and (2.14). Again these are nonunique because they depend on γ. This procedure, which is known as Δ-Σ hedging because the ratios (a_t, c_t) were originally labeled (Δ_t, Σ_t), is usually unsatisfactory owing to the higher transaction costs and lesser liquidity associated with trading the second derivative. It is therefore an important problem to determine how to hedge as "best as possible" (according to some criteria) with just the stock. We discuss the hedging problem further in Chapter 7.

2.6 Implied Volatility as a Function of Moneyness

Another reason that implied volatility, introduced in Section 2.1, is a particularly useful measure of the performance of a stochastic volatility model is that implied volatility is a function of a European option contract's *moneyness,* the ratio K/x of its strike price to the current stock price, as we now show. Given any Markovian stochastic volatility model under which the stock price satisfies

$$dX_s = rX_s \, ds + \sigma_s X_s \, dW_s^\star \qquad (2.22)$$

under some risk-neutral pricing measure $\mathbb{P}^{\star(\gamma)}$, suppose it is now time t and define $\tilde{X} = X/x$, where $X_t = x$, the current stock price. Then $(\tilde{X}_s)_{s \geq t}$ satisfies the same stochastic differential equation (2.22), with initial value $\tilde{X}_t = 1$, neither of which depends on the number x. The call option price

$$C = I\!E^{\star(\gamma)}\{e^{-r(T-t)}(X_T - K)^+ \mid X_t = x, \sigma_t\}$$
$$= I\!E^{\star(\gamma)}\{e^{-r(T-t)}(x\tilde{X}_T - K)^+ \mid \tilde{X}_t = 1, \sigma_t\}$$
$$= K\,I\!E^{\star(\gamma)}\{((x/K)\tilde{X}_T - 1)^+ \mid \tilde{X}_t = 1, \sigma_t\}$$
$$= KQ_1(t, x/K; T)$$

for some function Q_1 depending on x/K but not on x and K separately. The implied volatility I is computed from

$$C = C_{BS}(t, x; K, I),$$

and from the Black–Scholes formula (1.37) we have

$$C_{BS}(t, x; K, I) = KQ_2(t, x/K; I)$$

for some function Q_2 also depending on x/K but not on x and K separately. From the relation

$$KQ_1(t, x/K; T) = KQ_2(t, x/K; I),$$

we see that I must be a function of moneyness K/x but not of K and x separately.

   This is useful because it tells us that we can obtain the implied volatility curve predicted by a stochastic volatility model by solving – numerically, for example – the partial differential equation (2.16) with terminal condition $h(x) = (x - K)^+$ (or $h(x) = (K - x)^+$ for a put) for a fixed strike price K. Then plotting the resulting implied volatilities as a function of moneyness for different starting values x gives the same curve as if we were varying K. A finite-difference algorithm usually gives the numerical solution of (2.16) for a range of current stock prices x, and the "extra" information can be used to generate the implied volatility curve.

2.7   Market Price of Volatility Risk and Data

From the previous section, we know that (2.20) is a possible no-arbitrage derivative pricing formula for any equivalent martingale measure $I\!P^{\star(\gamma)}$. Much research has investigated the range of possible prices in general settings: for example, the possible derivative prices often fill an interval, and in many cases this interval is quite broad. If the volatility process is unbounded, then the range of European call option prices given by (2.20) with $H = (X_T - K)^+$ is between the price of the stock and the intrinsic value of the contract,

$$(X_t - K)^+ \leq V_t \leq X_t,$$

and the extremal values are attained for some equivalent martingale measures. When volatility is assumed to be bounded with $\sigma_t \in [\sigma_{\min}, \sigma_{\max}]$, the bounds are

$$C_{BS}(\sigma_{min}) \le V_t \le C_{BS}(\sigma_{max}).$$

This information is not useful unless one is certain of tight bounds on volatility – that is, unless volatility is almost constant. Worst-case analyses of stochastic volatility models usually result in such bands, which are too wide for practical purposes.

The alternative approach, which we shall follow here, is that the market selects a unique equivalent martingale measure under which derivative contracts are priced. The value of the market's price of the volatility risk γ can be seen only in derivative prices, since γ does not feature in the real-world model for the stock price (2.7). Sometimes this viewpoint is called selecting an approximating complete market, but the term is misleading because even though discounted derivative prices are martingales with respect to the particular $I\!P^{\star(\gamma)}$, they cannot be replicated by stock and bond alone. They can, however, be *superreplicated*: for example, buying the stock at time $t < T$ and holding it until expiration superhedges a short call position because $X_T > (X_T - K)^+$; this may yield a profit but never a loss. This (trivial) strategy is very expensive (it costs $\$X_t$), and much recent research concerns finding the cheapest superhedging strategy.

When we estimate the parameters for our model, we could use econometric methods such as maximum likelihood or method of moments on historical stock-price time-series data to find (α, β, m, ρ) plus the present volatility in the model (2.7), *having chosen an invertible function f*. Then we would need some derivative data to estimate γ, assuming for instance that it is constant.

The common practice, called *cross-sectional fitting,* is to estimate all the parameters from derivative data, usually at-the-money European option prices (or a section of the observed implied volatility surface). This ignores the statistical basis for the modeling but is easier to implement (especially when there is a formula for option prices under the chosen model) than time-series methods, which suffer because the (σ_t) process is not directly observable. If today is time $t = 0$ and we denote $\vartheta = (\alpha, \beta, m, \rho, \sigma_0, \gamma, \mu)$, then a typical least-squares fit is to observe call option prices $C^{obs}(K, T)$ for strike prices and expiration dates (K, T) in some set $\mathcal{K}$ and to solve

$$\min_{\vartheta} \sum_{(K,T) \in \mathcal{K}} (C(K, T; \vartheta) - C^{obs}(K, T))^2,$$

where $C(K, T; \vartheta)$ is the model-predicted call option price (either from solving the partial differential equation (2.16) with $h(x) = (x - K)^+$ or by Monte Carlo simulation). This process can be very slow and computationally intensive.

2.8   Special Case: Uncorrelated Volatility

It turns out that the $\rho = 0$ case is much easier to handle mathematically, and we present some results under this assumption. In equity markets it is widely believed that $\rho < 0$, but some studies suggest ρ is close to zero in foreign-exchange data.

The rest of the section (except for 2.8.3) assumes $\rho = 0$ and that (γ_t) is independent of the Brownian motion (W_t) driving the stock price, so that – under the risk-neutral probability $I\!P^{\star(\gamma)}$ – the volatility remains uncorrelated with (W_t). The results are valid for any volatility driving process (Y_t) (not necessarily OU, or even Itô).

2.8.1   Hull–White Formula

The pricing formula (2.20) can be simplified under our assumptions. We can condition on the path of the volatility process and, by iterated expectations, the price of a call option is given by

$$C(t, X_t, Y_t) = I\!E^{\star(\gamma)}\big\{ I\!E^{\star(\gamma)}\{e^{-r(T-t)}(X_T - K)^+ \mid \mathcal{F}_t, \sigma_s, t \le s \le T\} \mid \mathcal{F}_t \big\}.$$

The inner expectation is just the Black–Scholes computation with a time-dependent volatility, as in Section 2.2.1. The answer is the Black–Scholes formula (1.37) with appropriately averaged volatility, and so

$$C(t, x, y) = I\!E^{\star(\gamma)}\big\{ C_{BS}(t, x; K, T; \sqrt{\overline{\sigma^2}}) \mid Y_t = y \big\}, \qquad (2.23)$$

where

$$\overline{\sigma^2} = \frac{1}{T-t} \int_t^T f(Y_s)^2 \, ds \qquad (2.24)$$

and, for simplicity, we have assumed that (Y_t) is a Markov process. Thus, $\sqrt{\overline{\sigma^2}}$ is the root-mean-square time average of $\sigma(\cdot)$ over the remaining trajectory of each realization, and the call option price is the average over all possible volatility paths.

2.8.2   Stochastic Volatility Implies Smile

When the correlation is zero, it is known that stochastic volatility models predict European option prices whose implied volatilities smile. This result is an important success of these models, although the fact that one observes smile curves does not necessarily imply that volatility is stochastic. The derivation we present in this section is instructive because it relates the smile shape directly to the Black–Scholes lognormal model through which implied volatilities are calculated. In particular, it identifies the log-moneyness $\log(K/x)$ as an important quantity.

The full statement, due to Renault and Touzi, is as follows. *In a stochastic volatility model where $(\sigma_t)_{t\geq 0}$ and $(W_t)_{t\geq 0}$ are independent, suppose the risk premium process is a function of Y_t and t but not of X_t: $\gamma_t = \gamma(t, Y_t)$. Then, provided $\overline{\sigma^2}$ as defined by (2.24) is an L^2 random variable, the implied volatility curve $I(K)$ for fixed t, x, T is a smile – that is, it is locally convex around the minimum $K_{\min} = xe^{r(T-t)}$, which is the forward price of the stock.*

To see this, let us fix t and T and start with $\overline{\sigma^2}$ a Bernoulli random variable,

$$\overline{\sigma^2} = \begin{cases} \sigma_1^2 & \text{with probability } p, \\ \sigma_2^2 & \text{with probability } 1 - p, \end{cases}$$

under the measure $I\!\!P^{\star(\gamma)}$. Then, provided the listed assumptions hold, the Hull–White formula (2.23) applies:

$$C_{\mathrm{BS}}(K; I(p, K)) = pC_{\mathrm{BS}}(K; \sigma_1) + (1 - p)C_{\mathrm{BS}}(K; \sigma_2), \tag{2.25}$$

where $C_{\mathrm{BS}}(K; \sigma)$ denotes the standard Black–Scholes formula with only the K and σ arguments displayed. To simplify notation, hereafter we give only its volatility argument. We have written $I = I(p, K)$ to stress the dependence on p and K with x, t, and T fixed. Differentiating (2.25) with respect to K yields

$$\frac{\partial C_{\mathrm{BS}}}{\partial \sigma}(I(p, K))\frac{\partial I}{\partial K} + \frac{\partial C_{\mathrm{BS}}}{\partial K}(I(p, K)) = p\frac{\partial C_{\mathrm{BS}}}{\partial K}(\sigma_1) + (1 - p)\frac{\partial C_{\mathrm{BS}}}{\partial K}(\sigma_2),$$

so that

$$\mathrm{sign}\left(\frac{\partial I}{\partial K}\right) = \mathrm{sign}(g(p)),$$

where, for fixed K,

$$g(p) = p\frac{\partial C_{\mathrm{BS}}}{\partial K}(\sigma_1) + (1 - p)\frac{\partial C_{\mathrm{BS}}}{\partial K}(\sigma_2) - \frac{\partial C_{\mathrm{BS}}}{\partial K}(I(p, K))$$

because $\partial C_{\mathrm{BS}}/\partial \sigma > 0$ from the formula (2.2). Clearly, if $p = 0, 1$ then $I \equiv \sigma_2, \sigma_1$, respectively, so $g(0) = g(1) = 0$.

The key (and somewhat surprising) step is now to differentiate g with respect to p,

$$g'(p) = \frac{\partial C_{\mathrm{BS}}}{\partial K}(\sigma_1) - \frac{\partial C_{\mathrm{BS}}}{\partial K}(\sigma_2) - \frac{\partial^2 C_{\mathrm{BS}}}{\partial K \partial \sigma}(I(p, K))\frac{\partial I}{\partial p}; \tag{2.26}$$

we also differentiate (2.25) with respect to p,

$$\frac{\partial C_{\mathrm{BS}}}{\partial \sigma}(I(p, K))\frac{\partial I}{\partial p} = C_{\mathrm{BS}}(\sigma_1) - C_{\mathrm{BS}}(\sigma_2), \tag{2.27}$$

from which we can obtain an expression for $\partial I/\partial p$. Differentiating (2.26) and (2.27) again and substituting for $\partial I/\partial p$ and $\partial^2 I/\partial p^2$, we have

$$g''(p) = \frac{[C_{BS}(\sigma_1) - C_{BS}(\sigma_2)]^2}{\dfrac{\partial C_{BS}}{\partial \sigma}} \left(\frac{\partial^3 C_{BS}}{\partial K \partial \sigma^2} - \frac{\dfrac{\partial^2 C_{BS}}{\partial K \partial \sigma} \dfrac{\partial^2 C_{BS}}{\partial \sigma^2}}{\dfrac{\partial C_{BS}}{\partial \sigma}} \right),$$

where all the partial derivatives of C_{BS} are evaluated at I. Inserting directly from the Black–Scholes formula (1.37) then gives

$$g''(p) = 2\frac{[C_{BS}(\sigma_1) - C_{BS}(\sigma_2)]^2}{\dfrac{\partial C_{BS}}{\partial \sigma}(I)} \frac{\tilde{L}}{(T - t)I^3},$$

where $\tilde{L} = \log(xe^{r(T-t)}/K)$. We know from the calculation (2.2) that $\partial C_{BS}/\partial \sigma >$ 0 and so C_{BS} is increasing in the volatility parameter. Since $I > 0$, we conclude that $\text{sign}(g''(p)) = \text{sign}(\tilde{L})$. Thus, for $K < K_{\min} = xe^{r(T-t)}$, it follows that $\tilde{L} > 0$ and g can only achieve a minimum in $[0, 1]$, implying that $g < 0$ for $0 < p < 1$. Hence $\partial I/\partial K < 0$, and similarly for $K > K_{\min}$, $\partial I/\partial K > 0$, and $\partial I/\partial K = 0$ at $K = K_{\min}$. The implied volatility curve is locally convex around $K_{\min}$. This can be extended by induction on the number of possible values that σ^2 can take, as referenced in the notes.

This result says that stochastic volatility European option prices produce the smile curve for *any* volatility process uncorrelated with the Brownian motion driving the price process, and this robustness to specific modeling of the volatility is an important asset of these models.

2.8.3 Remark on Correlated Volatility

In general, the situation is more complicated when volatility is correlated with the Brownian motion (W_t) driving the stock price. To illustrate, we present a generalization of the Hull–White formula in the case of stochastic volatility driven by a Brownian motion correlated with (W_t). Again, for simplicity, we work with the OU-driven model; under an equivalent martingale measure $\mathbb{P}^{\star(\gamma)}$, the model is given by (2.18)–(2.19), which can be rewritten as

$$\frac{dX_t}{X_t} = r\,dt + \sigma_t\left(\sqrt{1 - \rho^2}\,d\hat{W}_t^\star + \rho\,d\hat{Z}_t^\star\right), \tag{2.28}$$

$$dY_t = \left[\alpha(m - Y_t) - \beta\left(\rho\frac{(\mu - r)}{f(Y_t)} + \gamma_t\sqrt{1 - \rho^2}\right)\right]dt + \beta\,d\hat{Z}_t^\star, \tag{2.29}$$

where $(\hat{W}_t^\star)$ and $(\hat{Z}_t^\star)$ are independent standard Brownian motions and $\sigma_t = f(Y_t)$. In fact, $\hat{W}^\star = \sqrt{1 - \rho^2}\,W^\star - \rho Z^\star$, which is independent of $\hat{Z}^\star = \rho W^\star + \sqrt{1 - \rho^2}\,Z^\star$.

Then, by Itô's formula,

$$d \log X_t = (r - \tfrac{1}{2}\sigma_t^2)\, dt + \sigma_t \left(\sqrt{1 - \rho^2}\, d\hat{W}_t^* + \rho\, d\hat{Z}_t^*\right).$$

Now, conditioned on the path of the second Brownian $\hat{Z}^*$, $\log X_t$ is normally distributed with

$$\mathbb{E}^{\star(\gamma)}\{\log X_T \mid \mathcal{F}_t\} = \left(\log x + \rho \int_t^T \sigma_s\, d\hat{Z}_s^* - \frac{1}{2}\rho^2 \int_t^T \sigma_s^2\, ds\right)$$
$$+ \left(r - \frac{1}{2}\overline{\sigma_\rho^2}\right)(T - t),$$

$$\mathrm{var}^{\star(\gamma)}\{\log X_T \mid \mathcal{F}_t\} = \overline{\sigma_\rho^2}(T - t),$$

where we define

$$\overline{\sigma_\rho^2} = \frac{1}{T - t} \int_t^T (1 - \rho^2)\sigma_s^2\, ds.$$

Using iterated expectations as in Section 2.8.1, the price of a European call is

$$C(t, x, y) = \mathbb{E}^{\star(\gamma)}\left\{ C_{\mathrm{BS}}(t, x\xi_t; K, T; \sqrt{\overline{\sigma_\rho^2}}) \mid Y_t = y \right\}, \tag{2.30}$$

where

$$\xi_t = \exp\left(\rho \int_t^T \sigma_s\, d\hat{Z}_s^* - \frac{1}{2}\rho^2 \int_t^T \sigma_s^2\, ds\right).$$

In this form, the price is the average of Black–Scholes prices with different volatilities *and* different starting values (current stock price).

Using the identity

$$C_{\mathrm{BS}}(t, x\xi_t; K, T; \sigma) = \xi_t C_{\mathrm{BS}}(t, x; K\xi_t^{-1}; \sigma),$$

which is easily verified from the Black–Scholes formula (1.37), we can rewrite (2.30) as

$$C(t, x, y) = \mathbb{E}^{\star(\gamma)}\left\{ \xi_t C_{\mathrm{BS}}(t, x; K\xi_t^{-1}, T; \sqrt{\overline{\sigma_\rho^2}}) \mid Y_t = y \right\}. \tag{2.31}$$

This is an average of the Black–Scholes formula jointly in the volatility and strike price under an equivalent probability with Radon–Nikodym derivative ξ_t, according to Girsanov's theorem (Section 1.4.1).

This formula, in either of the forms (2.30) or (2.31), is of practical use for Monte Carlo simulation of prices in a correlated stochastic volatility model because only one Brownian path has to be generated. Unlike the uncorrelated case, it does not directly reveal any information about the implied volatility curve and so a generalization of the Renault–Touzi argument given in Section 2.8.2 is not possible. In fact, numerical computation of implied volatilities from a correlated stochastic

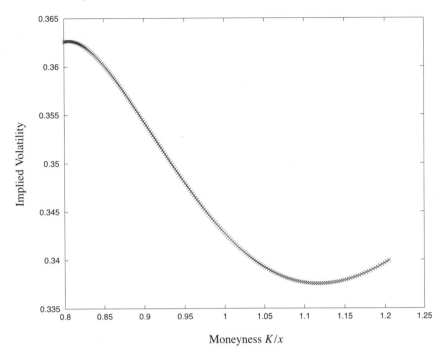

Figure 2.4. Implied volatilities predicted by the correlated stochastic volatility model (2.7), with $\rho = -0.2$, $f(y) = e^y$, and $\alpha = 1$, $\beta = \sqrt{2}$, $m = \log 0.1$, and $\gamma = 0$, $\mu - r = 0.16$ in (2.17). It is computed from a finite-difference solution to the partial differential equation (2.16) with terminal condition $h(x) = (x - K)^+$.

volatility model, such as shown in Figure 2.4, tells us that the convexity structure will be much more complicated. This illustrates the increase in complexity of these problems introduced by the skew effect.

2.9 Summary and Conclusions

We summarize the main features of the stochastic volatility approach. The positive aspects are:

- it directly models the observed random behavior of market volatility;
- it allows us to reproduce more realistic returns distributions, in particular tails that are fatter than lognormal;
- the asymmetry of the distribution is easily incorporated by correlating the noise sources; and

- the smile/smirk effect in option prices is exhibited in stochastic volatility models, where the correlation controls the skew.

However, new difficulties are associated with this approach.

- Volatility is not directly observed. As a consequence, estimation of the parameters of a specific model and the current level of volatility is not straightforward.
- There is no canonical stochastic volatility model that is generally accepted, and the relevance of explicit formulas for particular models is not obvious.
- We must deal with an incomplete market, which means that derivatives cannot be perfectly hedged with just the underlying asset. In addition, a volatility risk premium has to be estimated from option prices.

These problems can be addressed by exploiting the intrinsic time scale of volatility. This notion is explained in Chapter 3 and revealed in market data in Chapter 4. In Chapter 5, we develop an asymptotic theory based on this observation that yields a method with the following features.

- It applies to a large class of volatility processes that are driven by an ergodic process, and the results are not model-dependent.
- It incorporates a nonzero volatility risk premium and a nonzero correlation between volatility and asset price shocks that explains the much-observed skew or leverage effect.
- The asymptotic analysis yields a simple pricing and hedging theory that corrects the Black–Scholes theory to account for uncertain and changing volatility.
- The parameters needed for the theory are easily "read from the skew." That is, calibration from near-the-money European option–implied volatilities is simple and direct. Further, the theory does not require estimation of today's volatility level.

Notes

Empirical evidence for the smile curve of implied volatility appears in many studies, including Rubinstein (1985, using data from before the 1987 crash) and more recently Jackwerth and Rubinstein (1996), where the post-crash skew is documented.

Recent work on fitting an implied volatility surface, as discussed in Section 2.2, includes Rubinstein (1994), Dupire (1994), Derman and Kani (1994), and Avellaneda, Friedman, Holmes, and Samperi (1997). An extensive empirical study of the stability of the fitted surfaces can be found in Dumas, Fleming, and Whaley

(1998), and Lee (1999) presents an analysis of these results within a stochastic volatility framework.

Classical references investigating specific stochastic volatility models are Ball and Roma (1994), Heston (1993), Hull and White (1987), Scott (1987), Stein and Stein (1991), and Wiggins (1987). Good survey papers on the subject are Frey (1996), Ghysels, Harvey, and Renault (1996), and Hobson (1996).

Results and extensive references on superreplicating strategies in incomplete markets discussed in Section 2.7 are given in detail in Karatzas and Shreve (1998). The result of Section 2.8.2 appears in Renault and Touzi (1996) and the proof given here, including the induction step, is in Sircar and Papanicolaou (1999). The correlated Hull–White formula (2.30) of Section 2.8.3 is given by Willard (1996). It has been used recently by Lee (1999) to obtain convexity properties of the implied volatility for small ρ.

3 Scales in Mean-Reverting Stochastic Volatility

In this chapter we introduce the idea of *time scales* of the process that will drive volatility in the models discussed in Chapter 2. This is self-contained and uses basic facts of probability theory, which can be found in the textbook referenced in the notes at the end of the chapter. We will demonstrate that this is the right mathematical framework within which to use the observation that volatility is "bursty" in nature – the starting point of the asymptotic theory developed in Chapter 5.

The idea of scales is quite intuitive, and we will illustrate it here with pictures from numerical simulations. These will be used to show the effect of a fast scale in the volatility on stock prices, which, unlike volatility itself, is the observable quantity we deal with in practice. We conclude with a comparison with S&P 500 data.

3.1 Scaling in Simple Models

We introduce the property of burstiness or clustering, which is often observed in market volatility; we then show how this effect can be reproduced in models with a fast time scale.

Figure 3.1 shows two simulated realizations of possible volatility paths over the course of a year. In the first, volatility is low (4–8%) for a large part of that year (roughly $t = 0.1$ till $t = 0.8$ – over 8 months) and then is higher for the rest of the year. In the second path, volatility fluctuates between periods of high and periods of low far more often – it seems to be low for several days and then high for a similar period, then low again, and so on. That is, when it is low it often stays low for a period, and similarly when it is high.

Empirical studies and common experience suggest that the latter realization is a much more typical yearly volatility pattern than the former – it exhibits volatility clustering, or the tendency of high volatility to come in bursts. It is important

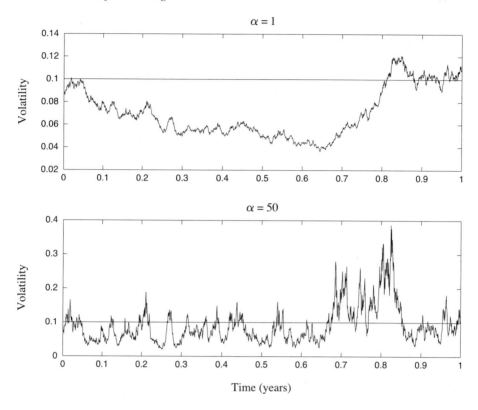

Figure 3.1. Simulated volatility paths. Note how volatility "clusters" in the latter case.

to realize that historical volatility is not directly observed but must be "filtered" or estimated from the observed price time series. Many empirical studies use *implied volatility,* defined in Section 2.1, as a proxy for the volatility process.

Burstiness is closely related to *mean reversion,* as introduced in Section 2.3.1, since a bursty process is returning to its mean, and the shorter the periods of the bursts, the more often it returns. This is the effect we want to model in the following examples.

3.2 Models of Clustering

Clustering, or mean reversion, is mathematically described by the concept of *ergodicity.* We shall not go fully into ergodic theory but restrict ourselves to simple Markovian examples. These are sufficient in practical applications and rich enough to exhibit the main features of volatility we want to describe. However, the theory we present in this book can all be repeated in a more general abstract setting, including high-dimensional non-Markovian processes driving the volatility model. Some examples are given in Chapter 10.

Recall from Section 2.3.1 that we model the volatility process (σ_t) as a positive function $\sigma_t = f(Y_t)$ of the *driving process* (Y_t). The following examples explain the relationship between burstiness and time scales in some simple models for (Y_t). They are chosen from the following three types of continuous-time Markov processes:

- continuous-time Markov chains on finite spaces;
- pure jump Markov processes in continuous spaces;
- diffusion processes.

In each case we explain the notions of infinitesimal generator, invariant distribution, and rate of mean reversion.

3.2.1   Example: Markov Chain

Suppose (Y_t) is a two-state Markov chain,

$$Y_t \in \{-1, 1\},$$

that represents a crude model of volatility switching between high and low states over time: $\sigma_t \in \{f(-1), f(1)\}$. The classical description of this Markov process is as follows. In an infinitesimal time dt, the process switches with probability $\alpha\, dt$ and remains in the same position with probability $1 - \alpha\, dt$. We assume here that α is a constant, meaning that (Y_t) is time-homogeneous and symmetric in the two states. If we define the 2×2 transition probability matrix $P(t)$

$$P(t) = \begin{pmatrix} I\!P\{Y_{s+t} = -1 \mid Y_s = -1\} & I\!P\{Y_{s+t} = 1 \mid Y_s = -1\} \\ I\!P\{Y_{s+t} = -1 \mid Y_s = 1\} & I\!P\{Y_{s+t} = 1 \mid Y_s = 1\} \end{pmatrix},$$

then this description implies that

$$P(\Delta t) = \begin{pmatrix} 1 - \alpha\Delta t & \alpha\Delta t \\ \alpha\Delta t & 1 - \alpha\Delta t \end{pmatrix} + o(\Delta t).$$

It follows that

$$P(t + \Delta t) - P(t) = (P(\Delta t) - I)P(t)$$

$$= \begin{pmatrix} -\alpha\Delta t & \alpha\Delta t \\ \alpha\Delta t & -\alpha\Delta t \end{pmatrix} P(t) + o(\Delta t),$$

so that, dividing by Δt and letting $\Delta t \downarrow 0$,

$$\frac{dP}{dt}(t) = \mathcal{L}P(t),\tag{3.1}$$

where we define the matrix

$$\mathcal{L} = \begin{pmatrix} -\alpha & \alpha \\ \alpha & -\alpha \end{pmatrix}.\tag{3.2}$$

In fact, $\mathcal{L}$ is exactly the infinitesimal generator of (Y_t) defined by

$$\mathcal{L}g(y) = \lim_{\Delta t \downarrow 0} \frac{I\!E\{g(Y_{\Delta t})\} - g(y)}{\Delta t}\tag{3.3}$$

for two-valued functions g on $\{-1, 1\}$. This is the same as the definition given in Section 1.5.1 for the generator of an Itô process; that it is given by $\mathcal{L}$ follows from

$$\frac{I\!E\{g(Y_{\Delta t})\} - g(y)}{\Delta t} = \frac{P(\Delta t) - P(0)}{\Delta t} g(y)$$

and the previous calculation.

In terms of expectations,

$$\frac{d}{dt} I\!E\{g(Y_t) \mid Y_0 = y\} = \frac{d}{dt}(P(t)g(y))$$

$$= \mathcal{L}P(t)g(y) = \mathcal{L}I\!E\{g(Y_t) \mid Y_0 = y\}.\tag{3.4}$$

From this we see that

$$I\!E\{g(Y_t) \mid Y_0 = y\} = e^{t\mathcal{L}}g(y),\tag{3.5}$$

where the exponential is the usual exponential of a 2×2 matrix.

It is a well-known fact that such a process stays in either state for a random holding time that is exponentially distributed. It then switches to the other state, stays there for another independent exponentially distributed random time, switches back, and so on. In fact, the parameter characterizing the exponential distribution is α. The common distribution of the positive holding times τ is given by the exponential distribution function

$$I\!P\{\tau \leq t\} = 1 - e^{-\alpha t}.$$

Notice that $1/\alpha$ can be thought of as the mean holding time:

$$I\!E\{\tau\} = 1/\alpha.$$

In other words, α is already a measure of the durations of the bursts of the process, since a large α means that the process is switching rapidly. We illustrate this in

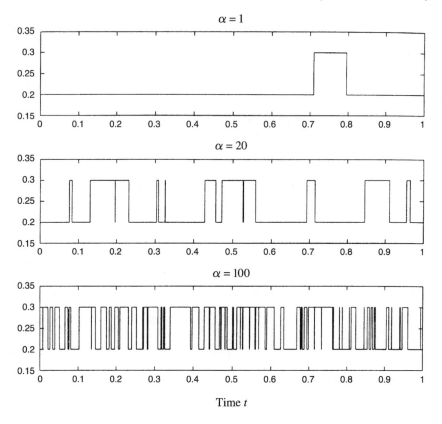

Figure 3.2. Simulated paths of $\sigma_t = f(Y_t)$, with (Y_t) a two-state Markov chain, showing the relation between clustering and the mean holding time $1/\alpha$.

Figure 3.2, which shows three typical realizations of $\sigma_t = f(Y_t)$ with $\alpha = 1$, $\alpha = 20$, and $\alpha = 100$.

We now introduce the fundamental concept of the *invariant distribution* of the process (Y_t) within this example: find an initial distribution for Y_0 such that, for any $t > 0$, Y_t has the same distribution. This is called the invariant distribution of the process, or its equilibrium state, because it does not change in time. We try to find the distribution of Y_0 such that

$$\frac{d}{dt} I\!E\{g(Y_t)\} = \frac{d}{dt} I\!E\{I\!E\{g(Y_t) \mid Y_0\}\} = 0,$$

where g is arbitrary. Parameterizing this two-valued distribution by the vector p_0 means we must solve

$$\frac{d}{dt} p_0^\star P(t)g = 0$$

for all two-dimensional vectors g, where $p_0^\star$ denotes the transpose of p_0. Using (3.1), we must have

$$p_0^\star \mathcal{L} P(t)g = 0$$

for all g, which implies

$$\mathcal{L}^\star p_0 = 0,$$

where $\mathcal{L}^\star$ is the adjoint operator of $\mathcal{L}$ and here is just the matrix transpose of (3.2). That is, the invariant distribution solves the adjoint equation, and this is a common feature of ergodic Markov processes. The unique solution is

$$I\!P\{Y_0 = -1\} = I\!P\{Y_0 = 1\} = \tfrac{1}{2},$$

independent of α.

We shall denote by $\langle \cdot \rangle$ the "integral" with respect to the invariant distribution:

$$\langle g \rangle = \tfrac{1}{2} g(-1) + \tfrac{1}{2} g(1).$$

In fact, the long-term behavior of (Y_t) is exactly described by the invariant distribution. To see this, we can compute the matrix exponential $e^{t\mathcal{L}}$ in (3.5) explicitly,

$$e^{t\mathcal{L}} = \frac{1}{2} \begin{pmatrix} 1 + e^{-2\alpha t} & 1 - e^{-2\alpha t} \\ 1 - e^{-2\alpha t} & 1 + e^{-2\alpha t} \end{pmatrix}, \tag{3.6}$$

and observe that

$$I\!E\{g(Y_t) \mid Y_0 = y\} \longrightarrow \langle g \rangle$$

as $t \to \infty$ and that the convergence is exponentially fast, with a rate proportional to α. That is, in the long run, the distribution of (Y_t) approaches its invariant distribution and forgets about where it started.

The formula (3.6) shows also that the distribution of Y_t depends only on the product αt, meaning that increasing α or speeding up the process have the same effect on the distribution. Under its invariant distribution, the process is fluctuating around its time-independent mean level at a rate proportional to α. We thus call α the **rate of mean reversion**. Notice that nowhere in this example are the fluctuations constrained to be small.

One additional feature about ergodic Markov processes that we illustrate with this example – and that we shall use repeatedly in the asymptotic calculations of Chapter 5 – concerns solutions of the homogeneous equation

$$\mathcal{L}\phi = 0. \tag{3.7}$$

Here ϕ is a vector in $I\!R^2$, and it is clear from the matrix (3.2) that any solution ϕ must have both its components equal. In other words, the equation (3.7) has only

constant solutions – that is, the null vectors of the generator of the Markov chain are constant (across the possible states).

3.2.2 Example: Another Jump Process

Our second example is a simple generalization of the first one. Here we allow the process (Y_t) to jump after exponentially distributed holding times to random values, uniformly distributed between -1 and 1. Jump sizes and holding times are independent, so (Y_t) is a pure jump Markov process in $[-1, 1]$. The parameter $1/\alpha$, the mean holding time, again measures the important intrinsic time scale of the process. This is illustrated in Figure 3.3.

In fact, the number of jumps N_t before time t is a Poisson process with intensity α:

$$IP\{N_t = k\} = \frac{(\alpha t)^k}{k!} e^{-\alpha t}$$

for nonnegative integers k. The density function p for the uniformly distributed jumps is simply

$$p(y) = \tfrac{1}{2}\mathbf{1}_{(-1,1)}(y).$$

Using definition (3.3) of the infinitesimal generator, we compute

$$IE\{g(Y_t)\} = IE\{g(Y_t) \mid N_t = 0\}IP\{N_t = 0\}$$
$$+ IE\{g(Y_t) \mid N_t \geq 1\}IP\{N_t \geq 1\}$$

for any bounded function g on $(-1, 1)$. For t small and $Y_0 = y$, we obtain

$$IE\{g(Y_t)\} = g(y)e^{-\alpha t} + \left(\int g(z)p(z)\,dz\right)\alpha t e^{-\alpha t} + \mathcal{O}(t^2),$$

where we have used $IP\{N_t = 0\} = e^{-\alpha t}$, $IP\{N_t = 1\} = \alpha t e^{-\alpha t}$, and $IP\{N_t \geq 2\} = \mathcal{O}(t^2)$. We deduce that

$$\frac{IE\{g(Y_t)\} - g(y)}{t} = \left(\int g(z)p(z)\,dz\right)\alpha e^{-\alpha t} - \left(\frac{1 - e^{-\alpha t}}{t}\right)g(y) + \mathcal{O}(t),$$

which gives

$$\mathcal{L}g(y) = \alpha \int (g(z) - g(y))p(z)\,dz$$

by taking the limit $t \downarrow 0$. From definition (3.3), $\mathcal{L}$ is the infinitesimal generator and is an integral operator in this case.

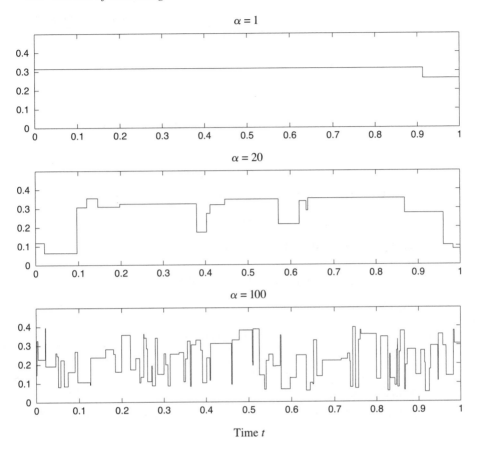

Figure 3.3. Simulated paths of $\sigma_t = f(Y_t)$, with (Y_t) a pure jump Markov process taking values in $(0.05, 0.4)$ for increasing mean-reversion rates α. The mean level $\langle f \rangle = 0.225$.

We now compute

$$
\frac{d}{dt} I\!E\{g(Y_t)\} = \lim_{\Delta t \downarrow 0} \frac{I\!E\{g(Y_{t+\Delta t})\} - I\!E\{g(Y_t)\}}{\Delta t}
$$

$$
= I\!E \left\{ \lim_{\Delta t \downarrow 0} \frac{I\!E\{g(Y_{t+\Delta t}) \mid Y_t\} - g(Y_t)}{\Delta t} \right\},
$$

where we have assumed we can interchange expectation and limit. Because the process (Y_t) is *time-homogeneous,* meaning that its statistics depend only on the time for which it has been running and not on the specific time, it follows that

$$\mathbb{E}\left\{ \lim_{\Delta t \downarrow 0} \frac{\mathbb{E}\{g(Y_{t+\Delta t}) \mid Y_t\} - g(Y_t)}{\Delta t} \right\} = \mathbb{E}\left\{ \lim_{\Delta t \downarrow 0} \frac{\mathbb{E}\{g(Y_{\Delta t}) \mid Y_0 = Y_t\} - g(Y_t)}{\Delta t} \right\}$$
$$= \mathbb{E}\{\mathcal{L}g(Y_t)\},$$

which gives

$$\frac{d}{dt}\mathbb{E}\{g(Y_t)\} = \mathbb{E}\{\mathcal{L}g(Y_t)\}.$$

To find the invariant distribution, we look for a density $p_0(y)$ for the initial value Y_0 such that

$$\frac{d}{dt}\mathbb{E}\{g(Y_t)\} = 0.$$

Using the iterated expectation $\mathbb{E}\{g(Y_t)\} = \mathbb{E}\{\mathbb{E}\{g(Y_t) \mid Y_0\}\}$, we want to solve

$$\mathbb{E}\{\mathcal{L}g(Y_0)\} = 0. \tag{3.8}$$

Consequently, we try to find $p_0(y)$ such that

$$\int p_0(y)\alpha\left(\int g(z)p(z)\,dz - g(y)\right)dy = 0$$

for all suitable g. In other words,

$$\mathcal{L}^* p_0 = \alpha(p - p_0) = 0.$$

It follows that

$$p_0(y) = p(y) = \tfrac{1}{2}\mathbf{1}_{(-1,1)}(y),$$

the uniform distribution.

As in the previous section, we denote by $\langle \cdot \rangle$ the expectation with respect to this distribution:

$$\langle g \rangle = \frac{1}{2}\int_{-1}^{1} g(z)\,dz.$$

A simple explicit computation shows that, under this invariant distribution,

$$\mathbb{E}\{g(Y_0)h(Y_t)\} = \langle g \rangle\langle h \rangle + e^{-\alpha t}(\langle gh \rangle - \langle g \rangle\langle h \rangle)$$

for any continuous bounded functions g and h. As $t \to \infty$, Y_t becomes independent of Y_0 and we say that the process *decorrelates* at the exponential rate α, which may also be interpreted as the rate of mean reversion of (Y_t).

In our two examples so far, the uniqueness of the invariant distribution and the decorrelation property are the main characteristics of an **ergodic process**. More precisely, the long-run time average of a bounded function g of an ergodic process is close to the statistical average with respect to its invariant distribution:

$$\lim_{t \to \infty} \frac{1}{t}\int_0^t g(Y_s)\,ds = \langle g \rangle.$$

This relation is true "almost surely" (a.s.), meaning on a set of probability 1. This is known as the ergodic theorem. It is a generalization of the classical law of large numbers for the sum of independent identically distributed (IID) random variables.

Recall from our previous example that the distribution of $f(Y_t)$ depends only on the product αt, and the same is true here. Allowing t to become large is the same *in distribution* as allowing the rate of mean reversion α to become large. Consequently, when mean reversion is fast,

$$\frac{1}{t} \int_0^t g(Y_s) \approx \langle g \rangle. \tag{3.9}$$

In particular, the mean-square time-averaged volatility $\overline{\sigma^2}$ is important in stochastic volatility models, as discussed in Section 2.8.1, so that

$$\overline{\sigma^2} = \frac{1}{T-t} \int_t^T f(Y_s)^2 \, ds \approx \langle f^2 \rangle, \tag{3.10}$$

a constant, when volatility is bursty. This will be the first key asymptotic calculation in Chapter 5.

3.2.3 *Example: Ornstein–Uhlenbeck Process*

As an example of a mean-reverting continuous diffusion process, let us now consider (Y_t) an Ornstein–Uhlenbeck process introduced in Section 2.3.1 as the solution of the stochastic differential equation (2.4):

$$dY_t = \alpha(m - Y_t) \, dt + \beta \, d\hat{Z}_t, \tag{3.11}$$

where $(\hat{Z}_t)$ is a standard Brownian motion. Recall that we wrote the explicit solution (2.5) for (Y_t), which showed that (Y_t) is a Gaussian process.

From Section 1.5.1 and formula (1.61), we know that the infinitesimal generator of the Markov process (Y_t) is the differential operator

$$\mathcal{L} = \alpha(m - y) \frac{\partial}{\partial y} + \frac{1}{2} \beta^2 \frac{\partial^2}{\partial^2 y}. \tag{3.12}$$

Again, as in (3.8), we look for an invariant distribution, which will be a distribution for Y_0 such that

$$\mathbb{E}\{\mathcal{L}g(Y_0)\} = 0 \tag{3.13}$$

for any smooth and bounded g. Let us denote the density function of this distribution by $\Phi(y)$, which satisfies

$$\int_{-\infty}^{\infty} \Phi(y)\mathcal{L}g(y) \, dy = 0,$$

by (3.13), for any g. Using integration by parts yields

$$\int_{-\infty}^{\infty} g(y)\mathcal{L}^{\star}\Phi(y)\,dy = 0,$$ (3.14)

where $\mathcal{L}^{\star}$, the *adjoint* of $\mathcal{L}$, is

$$\mathcal{L}^{\star} = -\alpha\frac{\partial}{\partial y}((m-y)\cdot) + \frac{1}{2}\beta^2\frac{\partial^2}{\partial^2 y}$$ (3.15)

and there are no boundary terms because $\Phi, \Phi' \to 0$ as $y \to \pm\infty$. If equation (3.14) holds for any smooth test function g, then Φ should satisfy

$$\mathcal{L}^{\star}\Phi = \tfrac{1}{2}\beta^2\Phi'' - \alpha((m-y)\Phi)' = 0,$$ (3.16)

where differentiations are with respect to the variable y. Solving this second-order ordinary differential equation with the constraint

$$\int_{-\infty}^{\infty}\Phi = 1$$

gives

$$\Phi(y) = \frac{1}{\sqrt{2\pi v^2}}\exp\left(-\frac{(y-m)^2}{2v^2}\right),$$

where

$$v^2 = \beta^2/(2\alpha).$$ (3.17)

This is exactly the density of the $\mathcal{N}(m, v^2)$ distribution, which was identified as the long-run distribution of this OU process in Section 2.3.1.

Under the invariant distribution, the covariance can be simply computed as

$$I\!E\{(Y_t - m)(Y_s - m)\} = v^2 e^{-\alpha|t-s|}.$$ (3.18)

We see again that the exponential rate of decorrelation of (Y_t) is proportional to α, and so $1/\alpha$ can be thought of as the typical **correlation time**. The parameter v^2 controls the size of the fluctuations, which we assume to be fixed as we consider larger and larger values of α. In other words, increasing α and keeping v fixed changes the degree of burstiness without affecting the *magnitude* of the fluctuations. This is to say that β increases as $\sqrt{\alpha}$, or $\beta = v\sqrt{2\alpha}$.

From (3.18) with $s = 0$, or from the solution of (3.11) which says that

$$Y_t \text{ is } \mathcal{N}\big(m + (y_0 - m)e^{-\alpha t}, \, v^2(1 - e^{-2\alpha t})\big)\text{-distributed,}$$

we see that with v fixed, the limits $t \to \infty$ and $\alpha \to \infty$ are the same in terms of the distributions. Consequently, the properties (3.9) and (3.10) noted in our previous example hold also for the OU process, where the average is with respect to the OU invariant distribution:

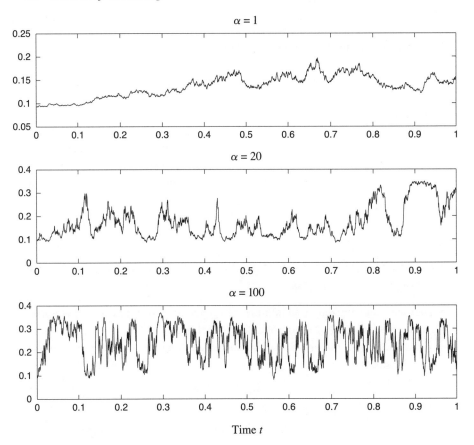

Figure 3.4. Simulated paths of $\sigma_t = f(Y_t)$, with (Y_t) a mean-reverting OU process defined by (3.11) and $f(y) = 0.35(\arctan y + \pi/2)/\pi + 0.05$, chosen so that $\sigma_t \in (0.05, 0.4)$. Notice how the mean-reversion rates α correspond to the duration of the bursts.

$$\langle g \rangle = \int_{-\infty}^{\infty} g(y)\Phi(y)\,dy.$$

In Figure 3.4 we illustrate with simulations the effect of α.

We also note that Figure 3.1 was simulated with an OU process for (Y_t) and the function $f(y) = e^y$. Remarkably, the dissimilar paths shown in this figure – and the three paths in Figure 3.4 – are each from the *same* process with different mean-reversion rates.

Finally within this example, we consider the null space of the generator $\mathcal{L}$, that is, solutions of

$$\mathcal{L}\phi = v^2\phi'' + (m - y)\phi' = 0. \tag{3.19}$$

Integrating this ordinary differential equation, we obtain

$$\phi(y) = c_1 \int_{-\infty}^{y} e^{(m-z)^2/2v^2} \, dz + c_2$$

for constants c_1 and c_2. We shall be interested in solutions that are "well-behaved" at infinity, and for this reason rapidly growing solutions associated with nonzero c_1 are excluded. Taking $c_1 = 0$ shows that the only admissible solutions of (3.19) are constant over the state y. This is exactly as we found with the Markov chain example (3.7) and a common property of generators of ergodic Markov processes.

3.2.4   Summary

We summarize the main facts about ergodic Markov processes that we shall use in the asymptotic calculations of Chapter 5.

- They are characterized by an infinitesimal generator $\mathcal{L}$ that may be, for example, a matrix, an integral operator, or a differential operator.
- The density p_0 of the invariant or stationary distribution of the process satisfies the adjoint equation

$$\mathcal{L}^* p_0 = 0.$$

- The homogeneous equation

$$\mathcal{L}\phi = 0$$

  has only constant solutions.
- The processes are associated with a characteristic mean reversion, holding, or correlation time α^{-1}. If α is large then, at finite times, the process is close to being under its invariant or long-run distribution.

3.3   Convergence to Black–Scholes under Fast Mean-Reverting Volatility

As we have seen in the examples of the last section, the time average of the square volatility approaches a constant as the rate of mean reversion α becomes large,

$$\overline{\sigma^2} = \frac{1}{T-t} \int_t^T f^2(Y_s) \, ds \longrightarrow \langle f^2 \rangle = \bar{\sigma}^2. \tag{3.20}$$

Notice the difference between the time average $\overline{\sigma^2}$, which is a random variable, and the expected square volatility $\bar{\sigma}^2$ under the invariant distribution.

We should expect that if mean reversion is very fast, the stochastic volatility price of a contract, discussed in Sections 2.4 and 2.5, should be close to its Black–Scholes price computed with the constant volatility $\bar{\sigma}$. We now show that this is true for the simpler uncorrelated case.

We present the analysis for a call option for simplicity. In this section, we assume the model (2.3), where $\sigma_t = f(Y_t)$ and where (Y_t) is any ergodic process (as in the examples described previously) independent of the Brownian motion (W_t).

Recall from the Hull–White formula derived in Section 2.8.1 that the call option price is given by

$$C(t, x, y) = I\!E^{\star(\gamma)}\big\{C_{BS}(t, x; \sqrt{\overline{\sigma^2}}) \mid Y_t = y\big\},$$

where $\overline{\sigma^2}$ is defined by the left-hand side of (3.20). In the large α limit, $\sqrt{\overline{\sigma^2}}$ can be replaced by $\bar{\sigma}$ given by the right-hand side of (3.20), and we have

$$\lim_{\alpha \to \infty} C(t, x, y) = C_{BS}(t, x; \bar{\sigma}).$$

Notice that the price does not depend on y, the value of the volatility driving process at time t.

We remark that if mean reversion is extremely fast then the constant-volatility Black–Scholes model is a good approximation. For fast but finite mean reversion, Black–Scholes needs to be corrected to account for random volatility; this will be the subject of Chapter 5.

In practice, we need to incorporate a skewness or leverage effect, which means a correlation between the volatility process and (W_t) in stochastic volatility models. Here the situation is much more complicated, as we have already mentioned in Section 2.8.3. Even though the convergence result is the same, the generalized Hull–White formula (2.30) does not give this directly. We will tackle this matter in Chapter 5, along with the correction.

3.4 Scales in the Returns Process

In reality, we do not observe the volatility process. The actual observable is the stock price, from which we can derive the returns process. We demonstrate from simulations that fast mean reversion can be identified qualitatively in the returns process. However, because α is the rate of mean reversion of the *hidden* volatility process, it is extremely difficult to estimate precisely from the simulated prices.

3.4.1 The Returns Process

If we rewrite the stochastic volatility model (2.3) as

$$\frac{dX_t}{X_t} - \mu \, dt = \sigma_t \, dW_t$$

then we may interpret the left-hand side as the "de-meaned" **returns process**, which (from the right-hand side) contains the pure fluctuation part of the price.

For simulation, we use a discrete-time version of the stochastic volatility model (2.3): the times are $\{t_n = n\Delta t, n = 0, \ldots, N\}$, where $\Delta t = T/N$ is the time spacing. We generate the discrete normalized returns process $\{D_n\}_{n=0}^{N}$ by

$$D_n = \sigma_n \varepsilon_n, \tag{3.21}$$

where $\{\sigma_n\}_{n=0}^{N}$ is obtained from the discrete version of our stochastic volatility models $f(Y_t)$ and $\{\varepsilon_n\}_{n=0}^{N}$ is a sequence of IID $\mathcal{N}(0, 1)$ random variables. The dt variance of dW_t is scaled out.

We present typical returns paths for a pure jump volatility process and for a continuous volatility process. For the first case we use our second model of Section 3.2.2; for the second case we use the OU model of Section 3.2.3. The first example in Section 3.2.1, volatility a two-state Markov chain, is essentially equivalent to the jump process example.

3.4.2 Returns Process with Jump Volatility

We denote by $\{\sigma_n\}_{n=0}^{N}$ the values $\{\sigma_{t_n}\}$ of the simulated jump process described in Section 3.2.2, sampled at the discrete times $\{t_n\}$. We do this for a small mean-reversion rate $\alpha = 1$ and a large rate $\alpha = 100$. The initial value for the volatility is chosen in both cases to be the mean volatility under the invariant distribution $\langle f \rangle$. The volatility process reverts slowly to its mean in the case $\alpha = 1$, so this choice has the effect of "fixing" the typical size of the fluctuations of the returns for our time of observation (a few months). In the case of α large, we know from the ergodic property described in Section 3.2.2 that the volatility process forgets about its starting value. Our choice of starting value is convenient for visually comparing the two cases, but a different choice would not change the difference in the structure of the fluctuations between them. We generate the returns process from (3.21) by using a sequence $\{\varepsilon_n\}$ that is independent of the volatility process. In other words, we deal with the uncorrelated case here.

The first column of Figure 3.5 shows a volatility path with $\alpha = 1$ in the top picture and the corresponding normalized returns path $\{D_n\}$ below it. The second

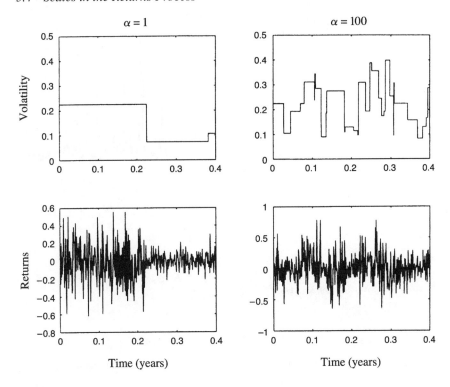

Figure 3.5. Simulated volatility and corresponding returns paths for small and large rates of mean reversion for the jump volatility model of Section 3.2.2.

column is the same for $\alpha = 100$. Looking at the bottom pictures only, we observe that the change in the size of the fluctuations is slow on the left. It is highly fluctuating up to about $t = 0.2$ and then exhibits smaller fluctuations for the rest of the time shown. On the right, the size of the fluctuations is changing a lot over the the same period of time. Since the *size of the fluctuations* is what we call volatility, fast mean reversion in volatility can be detected visually in the returns process. Even so, extracting the corresponding mean-reversion rate quantitatively from the returns data is extremely difficult; this is the subject of Chapter 4.

3.4.3   Returns Process with OU Volatility

For a volatility process driven by (Y_t) a mean-reverting OU process, as described in Section 3.2.3, we incorporate a correlation between volatility and stock-price

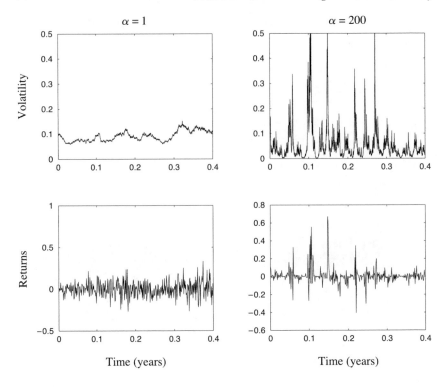

Figure 3.6. Simulated volatility and corresponding returns paths for small and large rates of mean reversion for the OU model of Section 3.2.3, $f(y) = e^y$.

shocks. This is simply done by discretizing the stochastic differential equation (3.11) – that is, writing the correlated Brownian motion $(\hat{Z}_t)$ as the sum of two independent Brownians, as in equation (2.6). We generate $\{Y_n\}_{n=0}^N$ by

$$Y_{n+1} = Y_n + \alpha(m - Y_n)\Delta t + \beta\sqrt{\Delta t}\left(\rho\varepsilon_n + \sqrt{1 - \rho^2}\eta_n\right),$$

where $\{\varepsilon_n\}$ and $\{\eta_n\}$ are independent sequences of IID $\mathcal{N}(0, 1)$ random variables. The returns are generated from (3.21) using the same sequence $\{\varepsilon_n\}$. To model a negative leverage or skew effect, we choose $\rho = -0.2$. Fixing the mean level m and variance $\nu^2 = \beta^2/(2\alpha)$ of the invariant distribution of the OU process driving the volatility leaves us to choose the rate of mean reversion α. We again consider a low value $\alpha = 1$ and a high value $\alpha = 200$. Typical paths are shown in Figure 3.6, with $\sigma_0 = \langle f \rangle$ for visual convenience as explained previously.

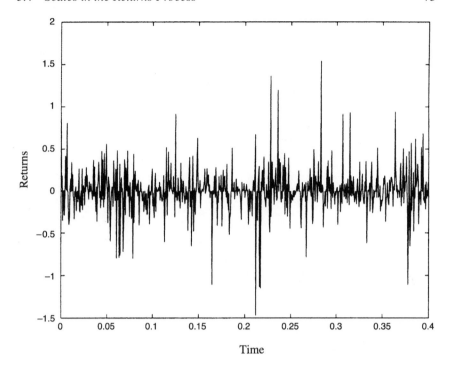

Figure 3.7. 1996 S&P 500 returns computed from half-hourly data.

Again we observe that the *size* of the fluctuation in the returns process of the bottom left picture is relatively constant over time, whereas in the large-α picture on the bottom right this size is changing a lot. This is exactly what we call fast mean-reverting stochastic volatility. The two volatility trajectories at the top of Figure 3.6 are from the same process with different α values, but the $\alpha = 200$ returns trajectory (bottom left) is not the $\alpha = 1$ returns trajectory (bottom right) speeded up.

3.4.4 S&P 500 Returns Process

In Figure 3.7 we show the S&P 500 returns process over the first five months of 1996. Real data will be analyzed more thoroughly in Chapter 4, but for a preliminary visual comparison we can see how the structure of the returns fluctuation size more resembles one from a *fast* mean-reverting stochastic volatility model than one from a model with slow mean reversion.

Notes

For more details on stochastic processes used in this chapter (the Markov chain, jump and OU processes) and their ergodic properties, we refer the reader to Breiman (1992). Time scaling of stochastic processes has been studied for years in many applications and often goes with asymptotic results; see the notes at the end of Chapter 5 for further references.

The half-hourly S&P 500 time series 1996 used in Figure 3.7 is from Olsen & Associates.

4 Tools for Estimating the Rate of Mean Reversion

The purpose of this chapter is to present statistical tools to estimate the rate of mean reversion of volatility from historical asset price data. We illustrate the methods using S&P 500 data and show that these data fit quite well the stochastic volatility models with fast mean reversion described in Chapter 3. Our estimate of the mean-reversion time of the S&P 500 volatility is about one and a half days. This is all that is needed for the reader to proceed to the asymptotic pricing and hedging theory presented in the rest of the book.

4.1 Model and Data

4.1.1 Mean-Reverting Stochastic Volatility

Recall the simplest class of mean-reverting or clustering stochastic volatility models in which volatility is modeled as a function of an ergodic process, for example, the OU process

$$dX_t = \mu X_t \, dt + f(Y_t) X_t \, dW_t \tag{4.1}$$

$$dY_t = \alpha(m - Y_t) \, dt + \beta\left(\rho \, dW_t + \sqrt{1 - \rho^2} \, dZ_t\right). \tag{4.2}$$

Here (Z_t) is another Brownian motion independent of (W_t), the source of additional randomness in the volatility fluctuations. The various parameters in (4.2) are

- the rate of mean reversion, α,
- the long-run mean m of Y_t,
- the volatility of the volatility, β, and

- the correlation coefficient $\rho \in (-1, 1)$ between the primary Brownian motion W_t and the secondary one $\rho W_t + \sqrt{1 - \rho^2} Z_t$ driving the volatility fluctuations.

Time is measured in years, so all rates are in annualized units.

Only the stock price or index (X_t) is observed, at discrete times; the fluctuating volatility $\sigma_t = f(Y_t)$ is not observed directly. Therefore, estimation of the parameters α, m, β, ρ is complicated since we are dealing with a *hidden* Markov model for the volatility.

Recall from Chapter 3 that by fast mean reversion we mean that the volatility driving process (Y_t) tends quickly to its equilibrium, which is a Gaussian with mean m and variance $\nu^2 = \beta^2/2\alpha$ as explained in Section 3.2.3. The equilibrium distribution is approached exponentially fast, with exponential rate α. We have fast mean reversion when α is large while the variance ν^2 is moderate. This means that β must be large as well, since $\nu^2 = \beta^2/2\alpha$. We want to estimate the rate of mean reversion, α.

4.1.2 Discrete Data

Typically with high-frequency data, tick-by-tick observations will be unevenly spaced with several thousand points per day. We average the data over five-minute intervals so that we have 72 data points per six-hour trading day. We collapse the time by eliminating overnights, weekends, and holidays. For the S&P 500 1994 data whose analysis is presented here, we have 251 trading days with $72 \times 251 = 18{,}072$ data points per year.

Let $\bar{X}_n$ denote the nth five-minute average of the asset price corresponding to the time $t_n = n\Delta t$, where Δt is five minutes. We introduce the *normalized fluctuation* of the data

$$\bar{D}_n = \frac{2(\bar{X}_n - \bar{X}_{n-1})}{\sqrt{\Delta t}(\bar{X}_n + \bar{X}_{n-1})}, \tag{4.3}$$

which is the increment $\bar{X}_n - \bar{X}_{n-1}$ divided by the average price $\frac{1}{2}(\bar{X}_n + \bar{X}_{n-1})$, normalized by $\sqrt{\Delta t}$. In other words, $\bar{D}_n$ is taken to be the observed realization of the asset price return.

It makes very little difference to the results if the five-minute averaging of the data is done on the X_n or on the normalized fluctuations. All of our analysis is based on the 18,072 data points for the normalized fluctuations $\{\bar{D}_n\}$ for 1994. The continuous analog of the normalized fluctuations can be written formally as

$$\frac{1}{\sqrt{\Delta t}} \frac{\Delta X_t}{X_t} = f(Y_t) \frac{\Delta W_t}{\sqrt{\Delta t}} + \mu\sqrt{\Delta t}, \tag{4.4}$$

based on the stochastic volatility model (4.1). The subtraction of the normalized mean return $\mu\sqrt{\Delta t}$ is omitted in the discrete normalized fluctuations $\{\bar{D}_n\}$ because it is negligibly small. Based on (4.4), we model the normalized fluctuation process by

$$\bar{D}_n = f(\bar{Y}_n)\varepsilon_n, \tag{4.5}$$

where $\{\varepsilon_n\}$ is a sequence of IID Gaussian random variables with mean 0 and variance 1 representing $\Delta W_t/\sqrt{\Delta t}$, and $\{\bar{Y}_n\}$ is a discrete-time equilibrium OU process representing (Y_{t_n}).

We will analyze the log absolute value of the normalized fluctuations

$$L_n = \log|\bar{D}_n| = \log(f(\bar{Y}_n)) + \log|\varepsilon_n|. \tag{4.6}$$

The effect of the log is to give us an *additive* white noise process.

Notice that with the choice $f(y) = e^y$, the first term on the right is the OU process itself, which has the equilibrium correlation function $I\!E\{\bar{Y}_{n+j}\bar{Y}_n\} = \nu^2 e^{-j\alpha\Delta t}$, as seen in (3.18). This exponential asymptotic decay in the equilibrium correlation function is true for general functions f:

$$I\!E\{f(\bar{Y}_{n+j})f(\bar{Y}_n)\} \sim \nu_f^2 e^{-j\alpha\Delta t}$$

as $\alpha \to \infty$, where ν_f^2 is the equilibrium variance of $\log(f(Y))$. This will be exploited to estimate the rate of mean reversion.

4.2   Variogram Analysis

We discuss the variogram approach to the estimation problem and display results from S&P 500 data.

4.2.1   Computation of the Variogram

We begin with a study of the empirical *structure function* or *variogram* of $\{L_n\}$:

$$V_j^N = \frac{1}{N}\sum_{n=1}^{N}(L_{n+j} - L_n)^2, \tag{4.7}$$

where j is the lag and N is the total number of points. This quantity is an estimator of $2c^2 + 2\nu_f^2(1 - e^{-j\alpha\Delta t})$, where $c^2 = \text{var}\{\log|\varepsilon|\}$ and we have assumed uncorrelated noises $\rho = 0$. This is because

$$IE\{(L_{n+j} - L_n)^2\}$$

$$= IE\{(L_j - L_0)^2\} \quad \text{(by stationarity)},$$

$$= IE\{(\log(f(\bar{Y}_j)) - \log(f(\bar{Y}_0)))^2\} + IE\{(\log(|\varepsilon_j|) - \log(|\varepsilon_0|)^2\},$$

$$= 2IE\{(\log(f(\bar{Y})))^2\} - 2IE\{\log(f(\bar{Y}_j))\log(f(\bar{Y}_0))\} + 2\,\text{var}\{\log(|\varepsilon|)\},$$

$$\approx 2v_f^2(1 - e^{-j\alpha\Delta t}) + 2c^2.$$

We discuss later the insensitivity of the results to the $\rho = 0$ assumption.

The top plot in Figure 4.1 shows the variogram (dotted line) of the log fluctuations $\{L_n = \log(|\bar{D}_n|)\}$ as a function of the lag for the 1994 S&P 500 after applying a ten-point median filter. According to our modeling (4.6), this is an exponentially correlated Markov process hidden in noise. The purpose of the median filter is to compensate for the singular noise $\log|\varepsilon_n|$.

There are three parameters to observe in the variogram: the intercept that gives the variance $2c^2$ of the noise $\log|\varepsilon_n|$; the distance between horizontal asymptote and the intercept that gives the variance $2v_f^2$ of the process $\log(f(\bar{Y}_n))$; and the curvature that determines α, the rate of mean reversion. To summarize, we have

$$V_j^N \approx 2c^2 + 2v_f^2(1 - e^{-\alpha|j|}), \tag{4.8}$$

ignoring the correlation (via ρ) between $\log|\varepsilon_n|$ and Y_n.

Note also in Figure 4.1 the *periodic component,* corresponding to day effects in the data; that is not in the model (4.5) but will be incorporated in what follows. The solid line is a fitted exponential obtained by nonlinear least-squares regression. The estimate was obtained by first computing the average of the empirical variogram over the first and last days and then, conditioned on these two parameters, carrying out a one-dimensional least-squares fit for α.

The estimated characteristic time of mean reversion
of S&P 500 volatility is 1.5 days.
(In the original units of calendar years, $1/\alpha$ is estimated to be 0.004.)

4.2.2   Comparison and Sensitivity Analysis with Simulated Data

We assess with numerical simulations the sensitivity of the estimates of α based on the variogram of the log of the normalized fluctuations $\bar{D}_n$.

Figure 4.2 shows our estimated standard deviation of the estimate for α. We simulate several realizations of the normalized fluctuations and estimate the rate

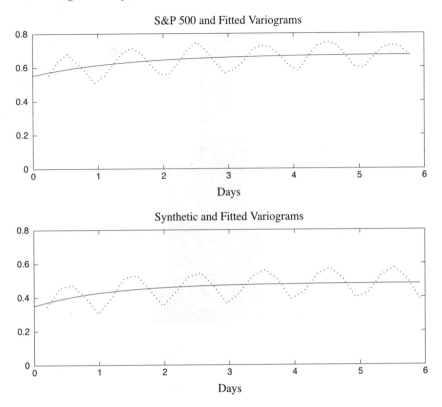

Figure 4.1. The dotted line in the top plot shows the empirical variogram for the S&P 500;
the solid line is the exponential fit from which the rate of mean reversion can be obtained.
The bottom plot is the same for simulated data based on our stochastic volatility model.

of mean reversion from the variogram by nonlinear least-squares fitting. The figure shows the distribution of the estimates when the rate of mean reversion in the
model is 1.5 days. In the next section, we discuss how we have incorporated the
observed periodic day effect in the simulations. Because of this, the estimator for
α based on an exponential model is slightly biased by about -0.2 days; in Figure 4.2, we have compensated for this systematic bias. The standard deviation for
the estimator of α is 0.4 days.

   The bottom plot in Figure 4.1 shows the estimate of the variogram for syn-
thetic log-normalized fluctuations, which have been median filtered (as were the
S&P 500 data). The solid line is the estimated exponential fit obtained by nonlinear least squares.

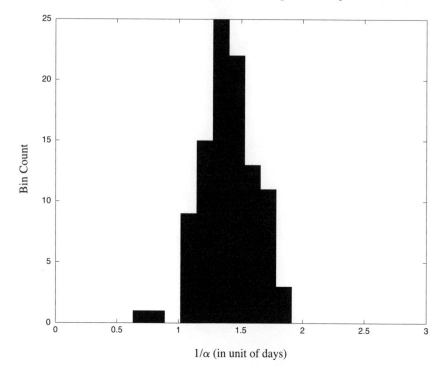

Figure 4.2. Empirical distribution of the estimator for the rate of mean reversion, α.

To generate simulated normalized fluctuations, we need to have estimates of the parameters m, α, β, and ρ. For the simulations we use the expOU model, where $f(y) = e^y$. We estimate the parameter $\bar{\sigma}^2 = \mathbb{E}\{\bar{D}_n^2\}$ by the mean square of the S&P 500 normalized fluctuations, $\bar{D}_n$; $\bar{\sigma}^2$ is related to m and $v_f^2 = v^2 = \beta^2/2\alpha$ by $\bar{\sigma}^2 = e^{2m+v^2}$ in the expOU case. We estimate v by observing that, if we assume at first that $\rho = 0$, then the (average) asymptote of the variogram in Figure 4.1 minus its intercept with the vertical axis is v^2, as in (4.8). Extensive simulations show that this estimate of v is insensitive to the correlation ρ in the range $[-0.8, 0]$; this is shown in Figure 4.3.

The parameter v could also have been estimated from the relation $\mathbb{E}\{D_n^4\} = 3\bar{\sigma}^4 e^{4v^2}$. However, this estimate depends on the fourth moment of the process $\{\varepsilon_n\}$ (as well as the Gaussianity of this process) and is sensitive to small deviations in its tails. The variogram analysis relies on the separation-of-scales assumption regarding the noise ε_n and $\exp(\bar{Y}_n)$. The estimation gives $\bar{\sigma} \approx 0.07$ and so from the variogram we have $v \approx 0.26$ and $1/\alpha \approx 1.5$ days.

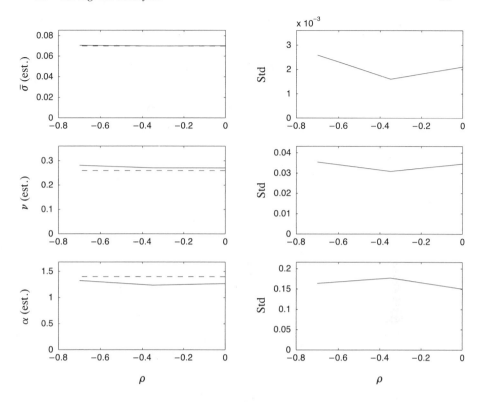

Figure 4.3. Insensitivity of the estimated parameters $\bar{\sigma}$, v, α to the correlation ρ in the range $[-0.8, 0]$. The dashed line is with simulated data.

4.2.3 The Day Effect

The day effect is put into the model by replacing (4.5) with

$$\bar{D}_n = f(\bar{Y}_n)g_n\varepsilon_n. \tag{4.9}$$

Here Y_n is the discrete OU process, ε_n a unit white noise process, and g_n a deterministic periodic function that models the systematic intraday variations in volatility. The estimate of g_n from the S&P 500 data is shown in Figure 4.4.

When the day effect is included in the simulations, the synthetic variogram in Figure 4.1 (bottom) is very close to that for the S&P 500 (Figure 4.1, top) except for the horizontal intercept. This intercept depends on the variance of the process $\log(|\varepsilon_n|)$, which in turn is sensitive to the fine properties in the tail of the distribution of ε. These are properties that, like the correlation ρ, we cannot expect to estimate precisely from data. The asymptotic analysis, which is the subject of the

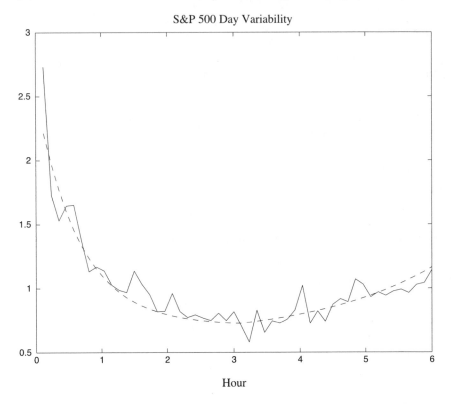

Figure 4.4. Estimated intraday variability envelope (dashed line). The solid line is the RMS of the observed normalized returns $\bar{D}_n$ as a function of trading hour, averaged over 250 days to give the variability envelope g_n. Note that the volatility is strongest at the beginning of the trading day.

rest of this book, does not require prior estimation of ρ but only that α be large. The effect of ρ for pricing and hedging derivatives will be derived from the observed implied volatility surface.

4.3   Spectral Analysis

Instead of using a variogram or structure function, we can analyze the empirical **spectrum** of the log fluctuations L_n given by (4.6). The analog of (4.8) in the frequency domain is obtained from the Fourier transform of the covariance of L_n. The spectrum is given by

$$\Gamma_\omega^N \approx c^2 + v_f^2 \hat{\Gamma}_f(\omega, \alpha), \tag{4.10}$$

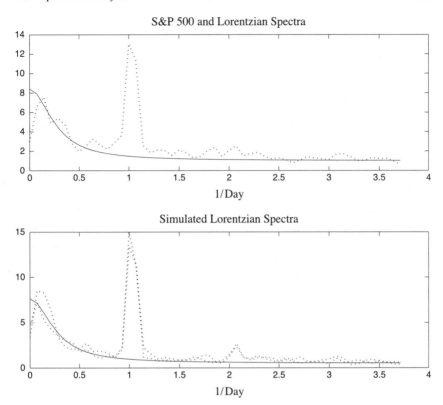

Figure 4.5. The top plot is the spectrum (dotted line) of $\log|\bar{D}_n|$ for the S&P 500 data; the Lorentzian model is shown by the solid line. Note the good match up to the day-effect pulse at one day. The bottom plot is the spectrum of two simulated realizations from the estimated model for $\bar{D}_n$, including the day effect g_j; the same Lorentzian model is shown by the solid line.

ignoring again the correlation between the noise $\log(|\varepsilon_n|)$ and the OU process Y_n as well as the day effect (which is just an additive impulse at the frequency corresponding to one day). The function $\nu_f^2\hat{\Gamma}_f(\omega, \alpha)$ is the spectrum of $\log(f(Y))$, where we show the dependency on the parameter α. In the expOU case, this is simply the Lorentzian

$$\nu_f^2\hat{\Gamma}_f(\omega, \alpha) = \frac{\nu^2}{\pi}\frac{\alpha}{\alpha^2 + \omega^2}.$$

In Figure 4.5, the dotted line in the top graph shows the spectrum for the S&P 500 data. The Lorentzian model is shown by the solid line. Note the good match up to the day effect, which is the impulse at the frequency of one day. The spectrum is computed using standard MATLAB routines for spectral estimation. The bottom plot is the spectrum of two realizations from the estimated model. The solid line is the same Lorentzian model, which is qualitatively very similar to that of the real data at the top.

It is important to note that, even though the Lorentzian spectrum is clearly visible for small frequencies, the variability in the spectrum in this frequency range is fairly high. For this reason we use the variogram rather than the spectrum to estimate the rate of mean reversion.

Notes

The data analysis presented here is in collaboration with Knut Solna. In Fouque, Papanicolaou, Sircar, and Solna (1999) we give more extensive results that include sensitivity analyses with respect to median filtering of the data, five-minute averaging, and length of averaging interval.

The high-frequency S&P 500 data comes from the Berkeley Options Database.

5 Asymptotics for Pricing European Derivatives

5.1 Preliminaries

In Chapter 2 we explained how stochastic volatility can be modeled as a function $\sigma_t = f(Y_t)$ of an auxiliary driving process (Y_t). In Chapter 3 we showed the effect of fast mean reversion in this process. In particular we know that, along with the invariant distribution of the process Y, three parameters are playing an important role in this effect.

(1) The *effective volatility* $\bar{\sigma}$ defined by

$$\bar{\sigma}^2 = \langle f^2 \rangle, \tag{5.1}$$

where the brackets denote the average with respect to the invariant distribution of Y (introduced in Section 3.2).

(2) The rate of mean reversion α or its inverse, the typical correlation time of (Y_t), which we will consider as a small quantity denoted by

$$\varepsilon = 1/\alpha, \tag{5.2}$$

as demonstrated in the previous chapter. This scaling models volatility clustering.

(3) The variance of the invariant distribution of Y, denoted by ν^2, which controls the long-run size of the volatility fluctuations. We assume that this quantity remains fixed as we consider smaller and smaller values of ε.

For simplicity, we do not include the daily periodic component described in Section 4.2.3, which is on the time scale ε. We will see in Section 10.2 that this component can be incorporated in the asymptotic analysis of this chapter without changing the nature of the results.

5.1.1 The Rescaled Stochastic Volatility Model

For clarity of exposition we first use the OU model introduced in Section 2.3.1. The asymptotic results we discuss do not depend on this choice, as we will discuss in Sections 5.2.4 and 10.3.

In the OU case the variance ν^2 is given by (3.17), which in terms of ε implies

$$\beta = \frac{\nu\sqrt{2}}{\sqrt{\varepsilon}}. \tag{5.3}$$

The OU stochastic volatility model (X, Y) described in Section 2.5 by (2.18) and (2.19) can be rewritten, under the risk-neutral probability $I\!P^{\star(\gamma)}$, in terms of the small parameter ε:

$$dX_t^\varepsilon = rX_t^\varepsilon\, dt + f(Y_t^\varepsilon)X_t^\varepsilon\, dW_t^\star, \tag{5.4}$$

$$d_t Y^\varepsilon = \left[\frac{1}{\varepsilon}(m - Y_t^\varepsilon) - \frac{\nu\sqrt{2}}{\sqrt{\varepsilon}}\Lambda(Y_t^\varepsilon)\right]dt + \frac{\nu\sqrt{2}}{\sqrt{\varepsilon}}\,d\hat{Z}_t^\star, \tag{5.5}$$

where we write $(X^\varepsilon, Y^\varepsilon)$ to indicate explicitly the dependence upon ε.

The function Λ is given, as in (2.17), by

$$\Lambda(y) = \rho\frac{(\mu - r)}{f(y)} + \gamma(y)\sqrt{1 - \rho^2}, \tag{5.6}$$

where we assume that the market price of volatility risk γ is a bounded function of y alone. If this assumption is not valid, it will be apparent a posteriori when the results are fitted to data in Chapter 6.

We also write

$$\hat{Z}_t^\star = \rho W_t^\star + \sqrt{1 - \rho^2}Z_t^\star \tag{5.7}$$

with $|\rho| < 1$, where $W^\star$ and $Z^\star$ are two independent standard Brownian motions under $I\!P^{\star(\gamma)}$.

5.1.2 The Rescaled Pricing Equation

We consider a European derivative given by its nonnegative payoff function $h(x)$ and its maturity time T. The price at time $t < T$ of this derivative is a function of the stock's present value $X_t^\varepsilon = x$ and the present value $Y_t^\varepsilon = y$ of the process driving the volatility. We denote this price by $P^\varepsilon(t, x, y)$, and we know from Section 2.5 that

$$P^\varepsilon(t, x, y) = I\!E^{\star(\gamma)}\{e^{-r(T-t)}h(X_T^\varepsilon) \mid X_t^\varepsilon = x,\ Y_t^\varepsilon = y\}. \tag{5.8}$$

Substituting our rescaled parameters α and β, given by (5.2) and (5.3), in the pricing partial differential equation (2.16), we can write

$$\frac{\partial P^\varepsilon}{\partial t} + \frac{1}{2} f(y)^2 x^2 \frac{\partial^2 P^\varepsilon}{\partial x^2} + \frac{\rho v \sqrt{2}}{\sqrt{\varepsilon}} x f(y) \frac{\partial^2 P^\varepsilon}{\partial x \partial y} + \frac{v^2}{\varepsilon} \frac{\partial^2 P^\varepsilon}{\partial y^2}$$

$$+ r\left(x \frac{\partial P^\varepsilon}{\partial x} - P^\varepsilon \right) + \left[\frac{1}{\varepsilon}(m - y) - \frac{v\sqrt{2}}{\sqrt{\varepsilon}} \Lambda(y) \right] \frac{\partial P^\varepsilon}{\partial y} = 0, \tag{5.9}$$

which has to be solved for $t < T$ with the terminal condition

$$P^\varepsilon(T, x, y) = h(x). \tag{5.10}$$

Observe again that the price of the derivative depends on the current level of volatility, which is not directly observable, and on the market price of volatility risk, which does not appear in the history of the stock price.

5.1.3   *The Operator Notation*

The partial differential equation (5.9) involves terms of order $1/\varepsilon$, $1/\sqrt{\varepsilon}$, and 1. In order to account for these three different orders, we introduce the following convenient notation:

$$\mathcal{L}_0 = v^2 \frac{\partial^2}{\partial y^2} + (m - y) \frac{\partial}{\partial y}, \tag{5.11}$$

$$\mathcal{L}_1 = \sqrt{2}\rho v x f(y) \frac{\partial^2}{\partial x \partial y} - \sqrt{2} v \Lambda(y) \frac{\partial}{\partial y}, \tag{5.12}$$

$$\mathcal{L}_2 = \frac{\partial}{\partial t} + \frac{1}{2} f(y)^2 x^2 \frac{\partial^2}{\partial x^2} + r\left(x \frac{\partial}{\partial x} - \cdot \right) = \mathcal{L}_{BS}(f(y)), \tag{5.13}$$

where

- $\alpha\mathcal{L}_0$ is the infinitesimal generator of the OU process Y given by (3.12);
- $\mathcal{L}_1$ contains the mixed partial derivative due to the correlation ρ between the two Brownian motions $W^\star$ and $\hat{Z}^\star$ (it also contains the first-order derivative with respect to y due to the market prices of risk); and
- $\mathcal{L}_2$ is the Black–Scholes operator (1.36) at the volatility level $f(y)$, also denoted by $\mathcal{L}_{BS}(f(y))$.

With this notation, the pricing partial differential equation (5.9) becomes

$$\left(\frac{1}{\varepsilon} \mathcal{L}_0 + \frac{1}{\sqrt{\varepsilon}} \mathcal{L}_1 + \mathcal{L}_2 \right) P^\varepsilon = 0, \tag{5.14}$$

with the terminal condition (5.10). With the assumptions made on the coefficients, this equation has a unique solution P^ε for any given value of $\varepsilon > 0$. In particular, the positive function f is assumed to be bounded away from zero to avoid degeneracy in the diffusion. The vanishing coefficient x^2 of the $\partial^2 P^\varepsilon / \partial x^2$ term might cause a problem but, as explained in Section 1.5.3, one can always perform a change of variable to $\log x$ to get around that difficulty.

The problem (5.14) is called a *singular perturbation* problem owing to the "diverging terms" when ε goes zero, keeping the time derivative in $\mathcal{L}_2$ of order 1. Nevertheless, as we will see, the solution P^ε has a limit as ε goes to zero and, moreover, we are able to simply characterize the first correction for small but nonzero ε.

5.2 The Formal Expansion

The method is to expand the solution P^ε in powers of $\sqrt{\varepsilon}$,

$$P^\varepsilon = P_0 + \sqrt{\varepsilon} P_1 + \varepsilon P_2 + \varepsilon \sqrt{\varepsilon} P_3 + \cdots, \tag{5.15}$$

where $P_0, P_1, \ldots$ are functions of (t, x, y) to be determined such that $P_0(T, x, y) = h(x)$. We are primarily interested in the first two terms $P_0 + \sqrt{\varepsilon} P_1$. The terminal condition for the second term is $P_1(T, x, y) = 0$.

Substituting (5.15) into (5.14) leads to

$$\frac{1}{\varepsilon} \mathcal{L}_0 P_0 + \frac{1}{\sqrt{\varepsilon}} (\mathcal{L}_0 P_1 + \mathcal{L}_1 P_0)$$
$$+ (\mathcal{L}_0 P_2 + \mathcal{L}_1 P_1 + \mathcal{L}_2 P_0)$$
$$+ \sqrt{\varepsilon} (\mathcal{L}_0 P_3 + \mathcal{L}_1 P_2 + \mathcal{L}_2 P_1)$$
$$+ \cdots$$
$$= 0. \tag{5.16}$$

This equation could equivalently be multipled by ε to avoid the diverging terms. We write it as shown so that the time derivative in $\mathcal{L}_2 P_0$ appears as an order-1 term.

5.2.1 The Diverging Terms

Equating terms of order $1/\varepsilon$, we must have

$$\mathcal{L}_0 P_0 = 0. \tag{5.17}$$

The operator $\mathcal{L}_0$, given by (5.11), is the generator of an ergodic Markov process and acts only on the y variable; hence P_0 must be a constant with respect to that variable, which implies that

$$P_0 = P_0(t, x), \tag{5.18}$$

a function of (t, x) only. This was demonstrated in examples at the end of Sections 3.2.1 and 3.2.3.

Similarly, in order to eliminate the term in $1/\sqrt{\varepsilon}$ in (5.16), we must have

$$\mathcal{L}_0 P_1 + \mathcal{L}_1 P_0 = 0. \tag{5.19}$$

The operator $\mathcal{L}_1$ given by (5.12) takes derivatives with respect to y, and we therefore deduce from (5.18) that $\mathcal{L}_1 P_0 = 0$ and consequently that $\mathcal{L}_0 P_1 = 0$. Using the same argument as for (5.17), it is clear that

$$P_1 = P_1(t, x), \tag{5.20}$$

a function of (t, x) only. This implies in particular that the combination of the first two terms $P_0 + \sqrt{\varepsilon} P_1$ will not depend on the present volatility.

5.2.2   Poisson Equations

Having eliminated the diverging terms in (5.16), we can continue to eliminate terms of order 1, $\sqrt{\varepsilon}$, ε, and so on successively because we are interested in asymptotic approximations that are more and more accurate as the parameter ε goes to zero.

The order-1 terms give

$$\mathcal{L}_0 P_2 + \mathcal{L}_1 P_1 + \mathcal{L}_2 P_0 = 0.$$

From (5.20) we know that $\mathcal{L}_1 P_1 = 0$, so this equation reduces to

$$\mathcal{L}_0 P_2 + \mathcal{L}_2 P_0 = 0. \tag{5.21}$$

The variable x being fixed, $\mathcal{L}_2 P_0$ is a function of y since $\mathcal{L}_2$ involves $f(y)$. Focusing on the y dependence only, equation (5.21) is of the form

$$\mathcal{L}_0 \chi + g = 0, \tag{5.22}$$

which is known as a **Poisson equation** for $\chi(y)$ with respect to the operator $\mathcal{L}_0$ in the variable y. This equation does not have a solution unless the function $g(y)$ is centered with respect to the invariant distribution of the Markov process Y whose infinitesimal generator is $\mathcal{L}_0$.

As in Section 3.2.3, we denote the invariant distribution of Y by

$$\Phi(y) = \frac{1}{\sqrt{2\pi} v} e^{-(y-m)^2/2v^2}. \tag{5.23}$$

The **centering condition**

$$\langle g \rangle = \int g(y) \Phi(y) \, dy = 0 \tag{5.24}$$

is necessary for (5.22) to admit a solution, as can be seen from the following calculation. We average (5.22) with respect to the invariant distribution of (Y_t), integrate
by parts, and use the definition (3.15) of the adjoint operator $\mathcal{L}_0^*$ and its property
(3.16) $\mathcal{L}_0^* \Phi = 0$:

$$\langle g \rangle = -\langle \mathcal{L}_0 \chi \rangle$$

$$= -\int (\mathcal{L}_0 \chi(y)) \Phi(y) \, dy$$

$$= \int \chi(y)(\mathcal{L}_0^* \Phi(y)) \, dy$$

$$= 0,$$

where there are no boundary terms because Φ, $\Phi' \to 0$ as $y \to \pm\infty$.

In Section 3.2 we used examples to explain that the long-run distribution of the
process Y is its invariant distribution,

$$\lim_{t \to +\infty} I\!E\{g(Y_t) \mid Y_0 = y\} = \langle g \rangle.$$

This convergence is actually exponential and, if we assume the centering condition (5.24), a formal solution of (5.22) is given by the converging integral

$$\chi(y) = \int_0^{+\infty} I\!E\{g(Y_t) \mid Y_0 = y\} \, dt. \tag{5.25}$$

This can be checked by writing

$$\mathcal{L}_0 \int_0^{+\infty} I\!E\{g(Y_t) \mid Y_0 = y\} \, dt = \int_0^{+\infty} \mathcal{L}_0 I\!E\{g(Y_t) \mid Y_0 = y\} \, dt$$

$$= \int_0^{+\infty} \frac{d}{dt} I\!E\{g(Y_t) \mid Y_0 = y\} \, dt$$

$$= -I\!E\{g(Y_t) \mid Y_0 = y\}\big|_{t=0}$$

$$= -g(y),$$

where we have used the definition of the infinitesimal generator of a Markov
process (recalled in Section 1.5.1) and

$$\lim_{t \to \infty} I\!E\{g(Y_t) \mid Y_0 = y\} = \langle g \rangle = 0.$$

All the solutions of the Poisson equation (5.22) are obtained by adding constants
to (5.25), since two solutions differ by a constant.

An important property of solutions of the Poisson equation (5.22) is their bound-
edness or growth in $|y|$ given the boundedness or growth of $g(y)$. Specifically, if
$\langle g \rangle = 0$ and $|g(y)| \le C_1(1 + |y|^n)$ for some constant C_1 and integer n, then

$$|\chi(y)| \le C_2(1 + |y|^n) \tag{5.26}$$

for some other constant C_2 and $n \ne 0$. When $n = 0$, the bound is logarithmic:
$|\chi(y)| \le C_2(1 + \log(1 + |y|))$.

To see this, we note that the Poisson equation (5.22) is

$$\mathcal{L}_0 \chi = \frac{v^2}{\Phi}(\Phi\chi')' = -g, \tag{5.27}$$

so that

$$\chi'(y) = \frac{-1}{v^2 \Phi(y)} \int_{-\infty}^{y} g(z)\Phi(z)\,dz = \frac{1}{v^2 \Phi(y)} \int_{y}^{\infty} g(z)\Phi(z)\,dz.$$

The second equality comes from the zero-mean property of g. For $y \to +\infty$ we have

$$\left| \frac{1}{v^2 \Phi(y)} \int_{y}^{\infty} g(z)\Phi(z)\,dz \right| \le \frac{\tilde{C}}{\Phi(y)} \int_{y}^{\infty} \Phi(z) z^n \, dz \approx \tilde{C} y^{n-1}$$

for some constant $\tilde{C}$, using the bound on g and integration by parts. A similar estimate holds for $y \to -\infty$. Integrating this estimate with respect to y gives (5.26).

5.2.3   The Zero-Order Term

In our situation, the centering condition in equation (5.21) gives

$$\langle \mathcal{L}_2 P_0 \rangle = 0.$$

Since P_0 does not depend on y, this is $\langle \mathcal{L}_2 \rangle P_0 = 0$ and, from the definition (5.13) of $\mathcal{L}_2$, we deduce that $\langle \mathcal{L}_2 \rangle = \mathcal{L}_{BS}(\bar{\sigma})$ where the effective volatility $\bar{\sigma}$ is defined by (5.1). Therefore, *the zero-order term $P_0(t, x)$ is the solution of the Black–Scholes equation*

$$\mathcal{L}_{BS}(\bar{\sigma}) P_0 = 0, \tag{5.28}$$

with the terminal condition $P_0(T, x) = h(x)$.

As the centering condition is satisfied we can write

$$\mathcal{L}_2 P_0 = \mathcal{L}_2 P_0 - \langle \mathcal{L}_2 P_0 \rangle = \frac{1}{2}(f(y)^2 - \bar{\sigma}^2)x^2 \frac{\partial^2 P_0}{\partial x^2}.$$

The second-order correction P_2, solution of the Poisson equation (5.21), is then given by

$$
\begin{aligned}
P_2(t, x, y) &= -\frac{1}{2}\mathcal{L}_0^{-1}(f(y)^2 - \bar{\sigma}^2)x^2 \frac{\partial^2 P_0}{\partial x^2} \\
&= -\frac{1}{2}(\phi(y) + c(t, x))x^2 \frac{\partial^2 P_0}{\partial x^2},
\end{aligned}
\tag{5.29}
$$

where $\phi(y)$ is a solution of the Poisson equation

$$\mathcal{L}_0\phi = f(y)^2 - \langle f^2 \rangle \tag{5.30}$$

and $c(t, x)$ is a constant in y that may depend on (t, x).

Equation (5.30) is the second-order linear differential equation

$$v^2\phi'' + (m - y)\phi' = f(y)^2 - \langle f^2 \rangle.$$

The probability density $\Phi(y)$ of the $\mathcal{N}(m, v^2)$-invariant distribution is given by (5.23), and we may compute $\phi(y)$ as follows:

$$\frac{1}{\Phi}(\Phi\phi')' = -\frac{y - m}{v^2}\phi' + \phi''$$

$$= \frac{1}{v^2}(v^2\phi'' + (m - y)\phi')$$

$$= \frac{1}{v^2}(f^2 - \langle f^2 \rangle).$$

One integration gives

$$\Phi(y)\phi'(y) = \frac{1}{v^2}\int_{-\infty}^{y}(f(z)^2 - \langle f^2 \rangle)\Phi(z)\,dz,$$

where we have zero on both sides at $y = \pm\infty$. In an abbreviated notation, ϕ' is given by

$$\phi' = \frac{1}{v^2\Phi}\int_{-\infty}^{\cdot}(f^2 - \langle f^2 \rangle)\Phi. \tag{5.31}$$

By (5.26), if f^2 is bounded then ϕ is bounded by a linear function in $|y|$.

5.2.4 The First Correction

Having chosen $P_0(t, x)$ such that the first three terms in (5.16) are zero, the next-order term in $\sqrt{\varepsilon}$ must equal zero, so that

$$\mathcal{L}_0 P_3 + \mathcal{L}_1 P_2 + \mathcal{L}_2 P_1 = 0. \tag{5.32}$$

This is again a Poisson equation for P_3 with respect to $\mathcal{L}_0$, which requires the centering or solvability condition

$$\langle \mathcal{L}_1 P_2 + \mathcal{L}_2 P_1 \rangle = 0. \tag{5.33}$$

Using the computation (5.29) for P_2, the fact that P_1 does not depend on y, and $\langle \mathcal{L}_2 \rangle = \mathcal{L}_{BS}(\bar{\sigma})$, we deduce that

$$\mathcal{L}_{BS}(\bar{\sigma})P_1 = \frac{1}{2}\langle \mathcal{L}_1\phi(y)\rangle x^2 \frac{\partial^2 P_0}{\partial x^2}. \tag{5.34}$$

Notice that $\mathcal{L}_1 c = 0$, since $\mathcal{L}_1$ takes derivatives with respect to y and $c(t, x)$ is independent of y. Using (5.12), one can compute the operator

$$\langle \mathcal{L}_1 \phi(y) \cdot \rangle = \sqrt{2} \rho v \langle f(y) \phi'(y) \rangle x \frac{\partial}{\partial x} - \sqrt{2} v \langle \Lambda(y) \phi'(y) \rangle.$$

and finally derive the equation for $P_1(t, x)$:

$$\mathcal{L}_{BS}(\bar{\sigma}) P_1 = \frac{\sqrt{2}}{2} \rho v \langle f \phi' \rangle x^3 \frac{\partial^3 P_0}{\partial x^3}$$

$$+ \left(\sqrt{2} \rho v \langle f \phi' \rangle - \frac{\sqrt{2}}{2} v \langle \Lambda \phi' \rangle \right) x^2 \frac{\partial^2 P_0}{\partial x^2}, \tag{5.35}$$

with the terminal condition $P_1(T, x) = 0$.

At this stage it is convenient to introduce notation for the first (small) correction,

$$\widetilde{P}_1(t, x) = \sqrt{\varepsilon} P_1(t, x), \tag{5.36}$$

which is the solution of

$$\mathcal{L}_{BS}(\bar{\sigma}) \widetilde{P}_1 = H(t, x), \tag{5.37}$$

where we define the source term H by

$$H(t, x) = V_2 x^2 \frac{\partial^2 P_0}{\partial x^2} + V_3 x^3 \frac{\partial^3 P_0}{\partial x^3}; \tag{5.38}$$

V_2 and V_3 are two small coefficients, given in terms of $\alpha = 1/\varepsilon$ by

$$V_2 = \frac{v}{\sqrt{2\alpha}} (2\rho \langle f \phi' \rangle - \langle \Lambda \phi' \rangle), \tag{5.39}$$

$$V_3 = \frac{\rho v}{\sqrt{2\alpha}} \langle f \phi' \rangle. \tag{5.40}$$

The first correction satisfies the Black–Scholes equation (5.37) with a zero terminal condition and a small source term computed from derivatives of the leading term $P_0(t, x)$. It is explicitly given by

$$\widetilde{P}_1(t, x) = -(T - t) \left(V_2 x^2 \frac{\partial^2 P_0}{\partial x^2} + V_3 x^3 \frac{\partial^3 P_0}{\partial x^3} \right). \tag{5.41}$$

This can be checked from the identity

$$\mathcal{L}_{BS}(\bar{\sigma})(-(T - t)H) = H - (T - t)\mathcal{L}_{BS}(\bar{\sigma})H,$$

and the last term is zero because

$$\mathcal{L}_{BS}(\bar{\sigma}) \left(x^n \frac{\partial^n P_0}{\partial x^n} \right) = x^n \frac{\partial^n}{\partial x^n} \mathcal{L}_{BS}(\bar{\sigma}) P_0 = 0 \tag{5.42}$$

for any positive integer n. The last can also be seen by changing to logarithmic stock price, where the Black–Scholes equation and the source term have constant coefficients.

The corrected price is given explicitly by

$$P_0 - (T - t)\left(V_2 x^2 \frac{\partial^2 P_0}{\partial x^2} + V_3 x^3 \frac{\partial^3 P_0}{\partial x^3} \right), \qquad (5.43)$$

where P_0 is the Black–Scholes price with constant volatility $\bar{\sigma}$.

5.2.5   Universal Market Group Parameters

The coefficients V_2 and V_3 are nontrivial functions of the original model given by (5.39) and (5.40), but the key observation is the *universality* of formula (5.43): any stochastic volatility model – when driven by an ergodic diffusion process (Y_t) such that our Poisson equations admit well-behaved solutions – will lead to a first correction of this type that is entirely characterized by the coefficients V_2 and V_3.

We shall also use the compact notation $\mathcal{A}$ for the differential operator on the right side of (5.37). Putting all the steps together, we can write

$$\mathcal{L}_{BS}(\bar{\sigma})\widetilde{P}_1 = \frac{1}{\sqrt{\alpha}} \langle \mathcal{L}_1 \mathcal{L}_0^{-1} (\mathcal{L}_2 - \langle \mathcal{L}_2 \rangle) \rangle P_0. \qquad (5.44)$$

This may be simplified to

$$\mathcal{L}_{BS}(\bar{\sigma})\widetilde{P}_1 = \mathcal{A} P_0,$$

where we define

$$\mathcal{A} = V_2 x^2 \frac{\partial^2}{\partial x^2} + V_3 x^3 \frac{\partial^3}{\partial x^3}. \qquad (5.45)$$

We emphasize again that, in this approach, detailed expressions for V_2 and V_3 are not important. Nevertheless, in order to illustrate how these two fundamental quantities are related to the model parameters, one can compute them in the case of the OU model by using the definition of ϕ' given in (5.31) and integrating by parts:

$$\langle f\phi' \rangle = \left\langle \frac{f}{\nu^2 \Phi} \int_{-\infty}^{\cdot} (f^2 - \langle f^2 \rangle)\Phi \right\rangle$$

$$= \frac{1}{\nu^2} \int_{-\infty}^{+\infty} f \int_{-\infty}^{\cdot} (f^2 - \langle f^2 \rangle)\Phi$$

$$= -\frac{1}{\nu^2} \langle F(f^2 - \langle f^2 \rangle) \rangle,$$

where F denotes an antiderivative of f. Similarly, using (5.6) yields

$$\langle \Lambda \phi' \rangle = \rho(\mu - r)\left\langle \frac{\phi'}{f} \right\rangle + \sqrt{1-\rho^2}\langle \gamma \phi' \rangle$$

$$= -\frac{\rho(\mu - r)}{v^2}\langle \tilde{F}(f^2 - \langle f^2 \rangle)\rangle - \frac{\sqrt{1-\rho^2}}{v^2}\langle \Gamma(f^2 - \langle f^2 \rangle)\rangle,$$

where $\tilde{F}$ and Γ denote antiderivatives of $1/f$ and γ, respectively. Combining these formulas with (5.39) and (5.40), we can relate the Vs to the original model parameters as follows:

$$V_2 = \frac{1}{v\sqrt{2\alpha}}\langle[-2\rho F + \rho(\mu - r)\tilde{F} + \sqrt{1-\rho^2}\Gamma](f^2 - \langle f^2 \rangle)\rangle,$$

$$V_3 = \frac{-\rho}{v\sqrt{2\alpha}}\langle F(f^2 - \langle f^2 \rangle)\rangle.$$

With a particular choice of a function f (e.g., $f(y) = e^y$), the quantities V_2 and V_3 become explicit functions of the remaining model parameters $(\mu, m, v, \rho, \alpha)$. This could be used along with statistical estimates of (say) μ, v, m, and α to estimate γ and ρ from a calibration of V_2 and V_3, as explained in Section 5.3. Again we stress that this is not how the asymptotic result is meant to be used, since V_2 and V_3 along with $\bar{\sigma}$ will be shown to be sufficient for many pricing and hedging problems.

5.2.6   *Probabilistic Interpretation of the Source Term*

We have seen that P_0 and $\tilde{P}_1$ are obtained as solutions of Black–Scholes equations with the effective volatility $\bar{\sigma}$. From Chapter 1 we know that such solutions can be represented as expectations of functionals of the geometric Brownian motion $\bar{X}_t$ defined by

$$d\bar{X}_t = r\bar{X}_t\,dt + \bar{\sigma}\bar{X}_t\,d\bar{W}_t,$$

where $\bar{W}$ is a standard Brownian motion under the probability $\overline{I\!P}$.

The leading-order term P_0 is obtained by writing the Feynman–Kac formula associated to (5.28),

$$P_0(t, x) = \overline{I\!E}\{e^{-r(T-t)}h(\bar{X}_T) \mid \bar{X}_t = x\}. \tag{5.46}$$

The correction $\tilde{P}_1$ is obtained by writing the Feynman–Kac formula with the source term defined in (5.38) and a zero terminal condition associated to (5.37). This yields

$$\tilde{P}_1(t, x) = \overline{I\!E}\left\{-\int_t^T e^{-r(s-t)}H(s, \bar{X}_s)\,ds \mid \bar{X}_t = x\right\}. \tag{5.47}$$

Adding (5.46) and (5.47), we obtain the **corrected pricing formula**

$$(P_0 + \widetilde{P}_1)(t, x) = \overline{I\!E}\left\{ e^{-r(T-t)}h(\bar{X}_T) \right.$$

$$\left. - \int_t^T e^{-r(s-t)}H(s, \bar{X}_s)\, ds \mid \bar{X}_t = x \right\}. \quad (5.48)$$

The *path-dependent payment stream* $-H(t, x)$ (to the holder) is computed from the second- and third-order derivatives of the classical Black–Scholes price P_0. It may be positive or negative and accounts dynamically for volatility randomness in a robust, model-independent way; it also accounts for the market price of volatility risk effectively selected by the market. Once $\bar{\sigma}$ has been estimated from historical data, only the small parameters V_2 and V_3 need to be calibrated. This issue will be addressed in Section 5.3.

5.2.7   Put–Call Parity

We remark that put–call parity, discussed in Section 1.3.4, is preserved by the corrected put and call prices. Recall from (1.41) that

$$C_0(t, x) - P_0(t, x) = x - Ke^{-r(T-t)},$$

where we denote by C_0 and P_0 the Black–Scholes call and put prices (respectively) computed with the same volatility $\bar{\sigma}$. Similarly denoting their corrections by $\widetilde{C}_1(t, x)$ and $\widetilde{P}_1(t, x)$ and using the formula (5.43), we see that

$$\widetilde{C}_1 - \widetilde{P}_1 = -(T-t)\left(V_3 x^3 \frac{\partial^3}{\partial x^3} + V_2 x^2 \frac{\partial^2}{\partial x^2} \right)(C_0 - P_0)$$

$$= -(T-t)\left(V_3 x^3 \frac{\partial^3}{\partial x^3} + V_2 x^2 \frac{\partial^2}{\partial x^2} \right)(x - Ke^{-r(T-t)})$$

$$= 0.$$

As a result,

$$(C_0(t, x) + \widetilde{C}_1(t, x)) - (P_0(t, x) + \widetilde{P}_1(t, x)) = x - Ke^{-r(T-t)},$$

and put–call parity is preserved.

5.2.8   The Skew Effect

The two terms in the source term of equation (5.37) do not play the same role. The third-order derivative term $V_3 x^3 (\partial^3 P_0 / \partial x^3)$ vanishes with ρ, as can be seen from (5.40), while the second-order derivative term $V_2 x^2 (\partial^2 P_0 / \partial x^2)$ depends also on the market price of volatility risk γ, as can be seen from (5.39) and (5.6).

This difference can also be illustrated by writing an equation for the corrected price $P_0 + \widetilde{P}_1$. Adding equation (5.28) for P_0 and equation (5.37) for $\widetilde{P}_1$, we obtain

$$\mathcal{L}_{BS}(\bar{\sigma})(P_0 + \widetilde{P}_1) = V_2 x^2 \frac{\partial^2 P_0}{\partial x^2} + V_3 x^3 \frac{\partial^3 P_0}{\partial x^3},$$

with the terminal condition $(P_0 + \widetilde{P}_1)(T, x) = h(x)$. Introducing, for V_2 small enough, the *corrected effective volatility*

$$\tilde{\sigma} = \sqrt{\bar{\sigma}^2 - 2V_2},$$

this equation can be rewritten as

$$\mathcal{L}_{BS}(\tilde{\sigma})(P_0 + \widetilde{P}_1) = -V_2 x^2 \frac{\partial^2 \widetilde{P}_1}{\partial x^2} + V_3 x^3 \frac{\partial^3 P_0}{\partial x^3}.$$

The new source term $-V_2 x^2 (\partial^2 \widetilde{P}_1/\partial x^2)$ is of order ε, since V_2 and $\widetilde{P}_1$ are both of order $\sqrt{\varepsilon}$. It is therefore negligible compared to the source term $V_3 x^3 (\partial^3 P_0/\partial x^3)$ and so the corrected price $P_0 + \widetilde{P}_1$ has the same order of accuracy as the solution $\widetilde{P}$ of

$$\mathcal{L}_{BS}(\tilde{\sigma})\widetilde{P} = V_3 x^3 \frac{\partial^3 P_0}{\partial x^3}, \tag{5.49}$$

with the same terminal condition $\widetilde{P}(T, x) = h(x)$. This shows that the V_2 term is simply a volatility level correction. Indeed, the V_3 term is of a different nature due to the presence of the third derivative, which cannot be a part of an infinitesimal generator. In fact, another reason for calling this the "skew" effect is its relation to the third moment of stock price returns. The presence of a negative correlation ρ is seen in the fatter left tail of returns distributions and, as we shall see in Section 5.3, to a downward sloping implied volatility surface.

In the *uncorrelated* case $\rho = 0$, equation (5.49) becomes a pure Black–Scholes equation with the constant volatility $\tilde{\sigma}$, which includes a correction due to the market price of volatility risk. At this order of approximation, the "smile effect" will only be seen in the correlated case. We see that the part of the correction associated with V_2 involves a level shifting of the volatility with which derivatives are priced. As we shall see from data, typically $V_2 < 0$ and the corrected effective volatility $\tilde{\sigma}$ is higher than historical average volatility $\bar{\sigma}$. This is a correction for volatility being random – *heteroskedasticity*, as it is sometimes called. It is related to the fourth moment or kurtosis of stock-price returns being larger than one would expect if volatility were constant.

5.3   Implied Volatilities and Calibration

The goal of this section is to show how our corrected price, and in particular the parameters V_2 and V_3, are easily related to observed prices or implied volatilities.

Toward this end we use call options, and we assume that the results obtained previously for smooth payoffs are also valid for call payoffs $h(x) = (x - K)^+$. The mathematical justification is discussed in the next section.

We first compute the approximate price $P_0 + \widetilde{P}_1$ given by (5.43) in the case of call options. Using the notation of Section 2.1, the leading term P_0 is $C_{BS}(t, x; K, T; \bar{\sigma})$ given by the Black–Scholes formula (Section 1.3.3)

$$P_0 = C_{BS} = xN(d_1) - Ke^{-r(T-t)}N(d_2), \tag{5.50}$$

with

$$d_{1,2} = \frac{\log(x/K) + (r \pm \frac{1}{2}\bar{\sigma}^2)(T - t)}{\bar{\sigma}\sqrt{T - t}}. \tag{5.51}$$

Using the relations

$$\frac{\partial d_{1,2}}{\partial x} = \frac{1}{x\bar{\sigma}\sqrt{T - t}},$$

$$N'(d) = \frac{1}{\sqrt{2\pi}}e^{-d^2/2},$$

$$e^{-d_2^2/2} = e^{-d_1^2/2}\left(\frac{xe^{r(T-t)}}{K}\right),$$

one can easily derive the **Delta**,

$$\frac{\partial P_0}{\partial x} = N(d_1),$$

and the **Gamma**,

$$\frac{\partial^2 P_0}{\partial x^2} = \frac{e^{-d_1^2/2}}{x\bar{\sigma}\sqrt{2\pi(T - t)}}.$$

In order to compute the source term H in equation (5.38), we introduce a new "Greek" – defined as $\partial^3 P_0/\partial x^3$ and which we propose to call the **Epsilon**, since it is related to a small correction. We compute

$$\frac{\partial^3 P_0}{\partial x^3} = \frac{-e^{-d_1^2/2}}{x^2\bar{\sigma}\sqrt{2\pi(T - t)}}\left(1 + \frac{d_1}{\bar{\sigma}\sqrt{T - t}}\right) \tag{5.52}$$

and obtain

$$H(t, x) = \frac{xe^{-d_1^2/2}}{\bar{\sigma}\sqrt{2\pi(T - t)}}\left(V_2 - V_3 - \frac{V_3 d_1}{\bar{\sigma}\sqrt{T - t}}\right). \tag{5.53}$$

From the formula (5.43), the correction is then given by

$$\widetilde{P}_1(t, x) = -(T - t)H(t, x) = \frac{xe^{-d_1^2/2}}{\bar{\sigma}\sqrt{2\pi}}\left(V_3\frac{d_1}{\bar{\sigma}} + (V_3 - V_2)\sqrt{T - t}\right). \tag{5.54}$$

At this stage we could use this explicit formula to fit our two small parameters V_2 and V_3 to observed prices of call options for different strikes and maturities. It turns out that it is more convenient to use implied volatilities, introduced in Section 2.1. Recalling that the implied volatility I is given by

$$C_{BS}(t, x; K, T; I) = C^{observed}(K, T),$$

we can expand $I = \bar{\sigma} + \sqrt{\varepsilon}I_1 + \cdots$ on the left-hand side and use our approximated price on the right-hand side,

$$C_{BS}(t, x; K, T; \bar{\sigma}) + \sqrt{\varepsilon}I_1 \frac{\partial C_{BS}}{\partial \sigma}(t, x; K, T; \bar{\sigma}) + \cdots = P_0(t, x) + \tilde{P}_1(t, x) + \cdots;$$

this leads to

$$\sqrt{\varepsilon}I_1 = \tilde{P}_1(t, x)\left[\frac{\partial C_{BS}}{\partial \sigma}(t, x; K, T; \bar{\sigma})\right]^{-1}.$$

In other words, up to an error of order $\mathcal{O}(\varepsilon)$, the implied volatility is given by

$$I = \bar{\sigma} + \tilde{P}_1(t, x)\left[\frac{\partial C_{BS}}{\partial \sigma}(t, x; K, T; \bar{\sigma})\right]^{-1} + \mathcal{O}(1/\alpha).$$

Differentiating C_{BS} with respect to the volatility parameter yields the so-called **Vega**,

$$\frac{\partial C_{BS}}{\partial \sigma} = \frac{xe^{-d_1^2/2}\sqrt{T-t}}{\sqrt{2\pi}}.$$

Combined with (5.54), this gives

$$I = \bar{\sigma} + \frac{V_3 d_1}{\bar{\sigma}^2\sqrt{T-t}} + \frac{V_3 - V_2}{\bar{\sigma}} + \mathcal{O}(1/\alpha).$$

The implied volatility is also conveniently written as

$$I = \bar{\sigma} + \frac{V_3}{\bar{\sigma}^3}\left(r + \frac{3}{2}\bar{\sigma}^2\right) - \frac{V_2}{\bar{\sigma}} - \frac{V_3}{\bar{\sigma}^3}\left(\frac{\log(K/x)}{T-t}\right) + \mathcal{O}(1/\alpha),$$

which is, up to order $\mathcal{O}(1/\alpha)$, an *affine function* of the log-moneyness-to-maturity ratio (LMMR):

$$I = a\left[\frac{\log\left(\dfrac{\text{strike price}}{\text{asset price}}\right)}{\text{time to maturity}}\right] + b + \mathcal{O}(1/\alpha), \tag{5.55}$$

with

$$a = -\frac{V_3}{\bar{\sigma}^3},$$

$$b = \bar{\sigma} + \frac{V_3}{\bar{\sigma}^3}\left(r + \frac{3}{2}\bar{\sigma}^2\right) - \frac{V_2}{\bar{\sigma}}.$$

For *calibration* purposes, having estimated a and b from the observed implied volatility surface, the group parameters V_2 and V_3 are given by

$$V_2 = \bar{\sigma}((\bar{\sigma} - b) - a(r + \tfrac{3}{2}\bar{\sigma}^2)) \qquad (5.56)$$

$$V_3 = -a\bar{\sigma}^3. \qquad (5.57)$$

Calibration from data will be discussed in the next chapter. We can already see that an affine fit (5.55) of the implied volatility should return a small slope a and an intercept b that is near $\bar{\sigma}$ in order for V_2 and V_3 to be small.

Notice that the approximation to the implied volatility surface (I as a function of K and $T - t$) given by the asymptotics is a function of the composite variable LMMR:

$$\text{LMMR} = \frac{\log(K/x)}{T - t}.$$

The time evolution of the surface as t and x change is built into this variable, too. A typical predicted surface using a and b values estimated from S&P 500 options prices (as explained in Chapter 6) is shown in Figure 5.1.

Finally, we observe that when $K = x$, $I \approx b$, so that b can be thought of as approximately the at-the-money implied volatility.

5.4   Accuracy of the Approximation

Our goal in this section is to show that the approximation $P_0 + \widetilde{P}_1$ to the price P is of order ε, in the sense that

$$|P - (P_0 + \widetilde{P}_1)| \le \text{constant} \times \varepsilon$$

when h is smooth and bounded (the constant is independent of ε but may depend on y, the current state of the volatility driving process). In order to do this, one can use the expansion (5.15) up to order 3 in $\sqrt{\varepsilon}$ and write

$$P^{\varepsilon} = P_0 + \sqrt{\varepsilon}P_1 + \varepsilon P_2 + \varepsilon\sqrt{\varepsilon}P_3 - Z^{\varepsilon},$$

where $P = P^{\varepsilon}$, $\widetilde{P}_1 = \sqrt{\varepsilon}P_1$, and Z^{ε} is a remainder that depends on ε as well.

At the terminal time T, we have

$$Z^{\varepsilon}(T, x, y) = \varepsilon(P_2(T, x, y) + \sqrt{\varepsilon}P_3(T, x, y)), \qquad (5.58)$$

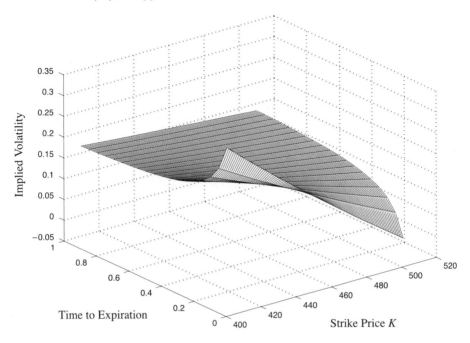

Figure 5.1. Typical implied volatility surface predicted by asymptotic analysis. It is linear in the composite variable LMMR with slope $a = -0.154$ and intercept $b = 0.149$ estimated from S&P 500 options data in Chapter 6. We take $t = 0$ and current asset price $x = 460$.

where we have used the terminal conditions $P(T, x, y) = P_0(T, x, y) = h(x)$ and $P_1(T, x, y) = 0$. This assumes smooth derivatives of P_0 in the domain $t \leq T$, which is justified when h is smooth. In particular, (5.58) shows that the terminal value of Z^ε is of order ε.

The next step is to compute $\mathcal{L}^\varepsilon Z^\varepsilon$ using the properties of $(P^\varepsilon, P_0, P_1, P_2, P_3)$:

$$
\begin{aligned}
\mathcal{L}^\varepsilon Z^\varepsilon &= \mathcal{L}^\varepsilon (P_0 + \sqrt{\varepsilon} P_1 + \varepsilon P_2 + \varepsilon \sqrt{\varepsilon} P_3 - P^\varepsilon) \\
&= \left(\frac{1}{\varepsilon} \mathcal{L}_0 + \frac{1}{\sqrt{\varepsilon}} \mathcal{L}_1 + \mathcal{L}_2 \right) (P_0 + \sqrt{\varepsilon} P_1 + \varepsilon P_2 + \varepsilon \sqrt{\varepsilon} P_3) - \mathcal{L}^\varepsilon P^\varepsilon \\
&= \frac{1}{\varepsilon} \mathcal{L}_0 P_0 + \frac{1}{\sqrt{\varepsilon}} (\mathcal{L}_0 P_1 + \mathcal{L}_1 P_0) \\
&\quad + (\mathcal{L}_0 P_2 + \mathcal{L}_1 P_1 + \mathcal{L}_2 P_0) + \sqrt{\varepsilon} (\mathcal{L}_0 P_3 + \mathcal{L}_1 P_2 + \mathcal{L}_2 P_1) \\
&\quad + \varepsilon (\mathcal{L}_1 P_3 + \mathcal{L}_2 P_2 + \sqrt{\varepsilon} \mathcal{L}_2 P_3) \\
&= \varepsilon (\mathcal{L}_1 P_3 + \mathcal{L}_2 P_2) + \varepsilon^{3/2} \mathcal{L}_2 P_3,
\end{aligned}
\tag{5.59}
$$

because P^ε solves the original equation $\mathcal{L}^\varepsilon P^\varepsilon = 0$ and P_0, P_1, P_2, P_3 have been chosen to cancel the first four terms.

Putting together $\mathcal{L}^\varepsilon Z^\varepsilon = \mathcal{O}(\varepsilon)$ and $Z^\varepsilon(T, x, y) = \mathcal{O}(\varepsilon)$ from (5.58), we obtain

$$Z^\varepsilon = \mathcal{O}(\varepsilon).$$

To see this, denote $\mathcal{L}^\varepsilon Z^\varepsilon = \varepsilon F^\varepsilon$ and $Z^\varepsilon(T, x, y) = \varepsilon G^\varepsilon$, where

$$F^\varepsilon(t, x, y) = \mathcal{L}_1 P_3(t, x, y) + \mathcal{L}_2 P_2(t, x, y) + \sqrt{\varepsilon}\mathcal{L}_2 P_3(t, x, y), \quad (5.60)$$

$$G^\varepsilon(x, y) = P_2(T, x, y) + \sqrt{\varepsilon}P_3(T, x, y). \quad (5.61)$$

We write the probabilistic interpretation of Z^ε:

$$Z^\varepsilon(t, x, y) = \varepsilon I\!\!E^\star \left\{ e^{-r(T-t)}G^\varepsilon(X_T^\varepsilon, Y_T^\varepsilon) \right.$$

$$\left. - \int_t^T e^{-r(s-t)}F^\varepsilon(s, X_s^\varepsilon, Y_s^\varepsilon)\,ds \mid X_t^\varepsilon = x, \; Y_t^\varepsilon = y \right\}. \quad (5.62)$$

Under the smoothness and boundedness assumption on the payoff function h and the boundedness of $f(y)$ and $\Lambda(y)$ – so that solutions of Poisson equations are at most linearly growing in $|y|$ by (5.26) – F^ε and G^ε are bounded uniformly in x and are at most linearly growing in $|y|$; the result follows. Consequently,

$$P(t, x, y) = (P_0(t, x) + \widetilde{P}_1(t, x)) + \mathcal{O}(\varepsilon),$$

which shows that the error in our approximation is of order $\varepsilon = 1/\alpha$.

The boundedness assumption on h can be easily removed by obtaining ε independent estimates for moments of X^ε (see the references given in the notes at the end of this chapter). The smoothness assumption cannot be removed by regularization of the payoff h, in the maximum norm. For European put or call options, for example, it is necessary to estimate the error Z^ε in a weak sense. In practice this means that the theory should not be used close to the strike time T, as we point out in the next section.

5.5   Region of Validity

The asymptotic approximations computed here and in other chapters cannot be expected to be valid for all (t, x) values. In particular, since the approximations involve averaging effects for the rapidly mean-reverting process (Y_t), they will not be as accurate close to the expiration date of a contract, when the process does not have sufficient time to fluctuate and the ergodic mean is a bad approximation. This is illustrated in Figure 5.2.

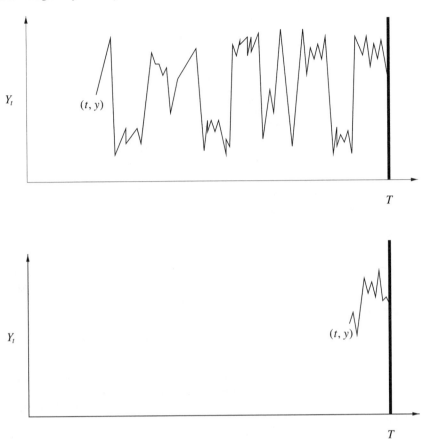

Figure 5.2. The top picture illustrates a path of (Y_t) starting well before the expiration date T. It has sufficient time to oscillate many times about its ergodic mean, so averaging approximations will be good in this case. In the bottom graphic, the process starts close to T and does not have time to mean revert. In this case, derivative prices would be sensitive to the starting value y.

For the example of European call options considered in Section 5.3, the implied volatility formula (5.55) will not be used when the time to expiration $T - t$ is small or when the option is far out of the money, that is, when $|\log(K/x)|$ is large because the supposedly small correction there becomes large. This is because we have divided the pricing correction by the Vega, which is small in these regimes.

One way to understand this is by recognizing that the asymptotic formulas are derived from time-scale limiting arguments that are similar to those underpinning

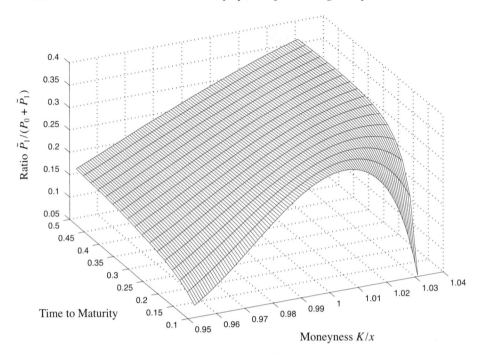

Figure 5.3. Ratio of correction $\widetilde{P}_1$ to corrected price $\widetilde{P}$ for a European call option using parameter values calibrated from the observed S&P 500 implied volatility surface: $a = -0.154$, $b = 0.149$ and $\bar{\sigma} = 0.1$, $r = 0.02$. These values give $V_2 = -0.0044$ and $V_3 = 0.000154$.

all continuous-time financial models. Just as a Brownian motion–based model can be a good approximation when the typical time between trades is small compared to the length of the derivative contract, so the approximations are good here as long as the volatility has had "enough" time to fluctuate.

That is, the correction is to be used *away* from the expiration date (or from the optimal exercise boundary in American-style option problems). For close-to-maturity contracts – where Black–Scholes performs poorly – irrational effects are prevalent, and these corrections are not remedial to that problem. It is doubtful whether diffusion models can be useful over those short time scales.

In Figure 5.3 we plot the ratio of the call correction $\widetilde{P}_1(t, x)$ in (5.54) to the corrected price $P_0(t, x) + \widetilde{P}_1(t, x)$ of (5.50) to illustrate the order of magnitude of the correction.

In the next chapter, we discuss the practical use of these asymptotic results.

Notes

The mathematical theory of the type of asymptotic analysis presented here can be found in Blankenship and Papanicolaou (1978), Papanicolaou, Stroock, and Varadhan (1977), or Papanicolaou (1978). It is also closely related to homogenization theory, which can be found, for example, in Chapter 3 of Bensoussan, Lions, and Papanicolaou (1978). The main new feature in the stochastic volatility context is that the first correction $\widetilde{P}_1$ does not depend on the current level y of the volatility driving process (Y_t) and can be computed explicitly. This calculation for European derivatives first appeared in Fouque, Papanicolaou, and Sircar (1998) and Fouque, Papanicolaou, and Sircar (2000a).

6 Implementation and Stability

In this chapter, we recapitulate the results of the asymptotic analysis of Chapter 5 and the idea of fast mean reversion. We discuss step-by-step how to implement the methodology and calibrate from the observed implied volatility surface. In Section 6.2, we demonstrate the stability over time of the calibration – a standard by which any approach should be judged.

6.1 Step-by-Step Procedure

In a stochastic volatility environment, the methodology introduced in the previous chapters is implemented as follows.

(1) Estimate $\bar{\sigma}$, the *effective historical volatility,* from stock-price returns.
(2) Use variogram analysis of historical stock-price returns to establish that volatility is *fast mean-reverting.*
(3) Fit an affine function in the composite variable called the *log-moneyness-to-maturity ratio* (LMMR) to the implied volatility surface I across strikes and maturities for liquid options:

$$I = a \left(\frac{\log\left(\dfrac{\text{strike price}}{\text{stock price}}\right)}{\text{time to maturity}} \right) + b. \tag{6.1}$$

(4) From the estimated slope a, the intercept b, and the effective volatility $\bar{\sigma}$, calculate the two fundamental quantities

$$V_2 = \bar{\sigma}((\bar{\sigma} - b) - a(r + \tfrac{3}{2}\bar{\sigma}^2)), \tag{6.2}$$

$$V_3 = -a\bar{\sigma}^3, \tag{6.3}$$

where the instantaneous interest r is assumed to be constant.

(5) The price of a European contract with date-T payoff function $h(X_T)$, corrected for stochastic volatility, is given by

$$P_0 - (T - t)\left(V_2 x^2 \frac{\partial^2 P_0}{\partial x^2} + V_3 x^3 \frac{\partial^3 P_0}{\partial x^3}\right),$$

where $P_0(t, x)$ is the Black–Scholes pricing function for that contract computed with constant volatility $\bar{\sigma}$ and interest r.

(6) We need only the market constants V_2 and V_3 to correct (for stochastic volatility) the Black–Scholes prices of other types of derivatives such as barrier, Asian, and American, as described in Chapters 8 and 9.

(7) These same market constants are used in the hedging strategies described in Chapter 7.

6.2   Comments about the Method

This approach to stochastic volatility market has three notable features: model independence, parsimony of parameters, and stability of parameter estimates.

Model Independence.   The approach just described requires that volatility be mean-reverting but does not otherwise depend in an essential way on how the volatility is modeled. We have seen in Chapter 4 how to infer from price data the fast mean reversion or clustering property of volatility, as demonstrated from S&P 500 data. No specific stochastic volatility model need be assumed to use these tools.

In addition, the correlation between stock price and volatility shocks cannot be seen in the variogram analysis and would be extremely difficult to estimate stably by other means. However, the presence of a skew in implied volatility signifies that this correlation is extremely important for derivative pricing and hedging. The V_2 and V_3 revealed by the asymptotics and estimated from the smile contain the essential information about the correlation and the market price of volatility risk for these problems.

We stress again that this method is *not* model-specific: it does not depend on choosing a particular model $\sigma_t = f(Y_t)$ or a particular ergodic driving diffusion (Y_t). Without choosing a specific (f, Y), the asymptotic analysis leads to three observable quantities $\bar{\sigma}, a, b$ (or equivalently $(\bar{\sigma}, V_2, V_3)$) with no need to estimate the specific volatility model parameters separately; only the Vs (which contain these) are needed. Nor is the present value Y_t required.

Parsimony of Parameters.   The method simplifies enormously the parameter estimation problem. Any stochastic volatility model will be described by at least

Table 6.1: *Parameters*

Model Parameters	Necessary Parameters
Mean level of volatility	Mean level of volatility $\bar{\sigma}$
Variance of volatility	Slope of implied volatility line-fit a
Rate of mean reversion of volatility	Intercept of implied volatility line-fit b
Correlation between shocks	
Volatility risk premium	

the five quantities listed in the left column of Table 6.1. The asymptotics reveal the group parameters V_2 and V_3 and their relation to the easily calibrated a, b, and $\bar{\sigma}$; they capture the essential effects of stochastic volatility for derivative management problems.

We can estimate a and b from the smile, the mean volatility $\bar{\sigma}$ from returns data, and the rate of mean reversion α and variance v_f^2 of the volatility from the variogram. In principle, we could then recover estimates of the correlation ρ and the volatility risk premium γ (if it is assumed constant) *but only by choosing a specific pair* (f, Y). However, one would not do so because (a) the first three are enough for the derivative problems we are interested in and (b) we would then be subject to the instabilities of a model-dependent approach.

Stability of Parameter Estimates.   We have tested a posteriori the feasibility of the theory-predicted LMMR line-fit for actual implied volatility data. The market in at- and near-the-money European options on the S&P 500 index is liquid, and we show in Figure 6.1 daily estimates of the slope and intercept coefficients $\hat{a}$ and $\hat{b}$ derived by fitting Black–Scholes implied volatilities from observed S&P 500 European call-option prices to (6.1) across strikes K and maturities T. We used contracts within 3% of the money,

$$|K/x - 1| \leq 3\%,$$

and at least three weeks to expiration. We estimated a and b by a least-squares fit on days when there were at least 100 datapoints (i.e., sufficient liquidity).

We observe from the results that the estimated slopes $\hat{a}$ and the estimated $\hat{b} - \bar{\sigma}$ are small. This strongly supports the fast mean-reverting hypothesis and validates use of the asymptotic formula, since the full skew formulas (5.39) and (5.40) show that these quantities are terms of order $1/\sqrt{\alpha}$. We also find that the estimates $\hat{a}$ and

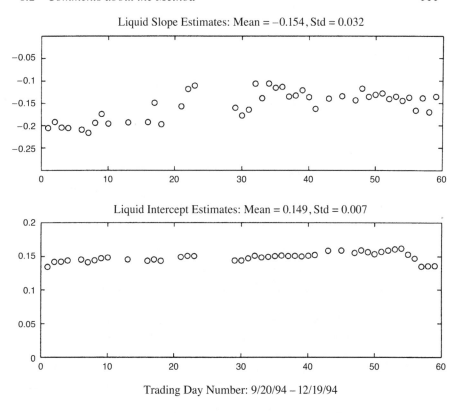

Figure 6.1. Daily fits of S&P 500 European call option implied volatilities to a straight line in LMMR, excluding days when there is insufficient liquidity (16 days out of 60).

$\hat{b}$ over a 60-day period are stable, as evidenced by the standard deviations of the daily estimates reported in Figure 6.1.

Our estimated parameter values,

$$a = -0.154, \quad b = 0.149, \quad \bar{\sigma} = 0.1,$$

along with the interest rate $r = 0.02$, imply via (6.2) and (6.3) the small values of the group parameters for the S&P 500:

$$V_2 = -0.0044, \qquad V_3 = 0.000154.$$

It is not entirely surprising that the parsimonious group parameters exhibit this stability in time, for through the asymptotic analysis they have been designed to capture the dynamics of price, the volatility, and its associated risk.

6.3   Dividends

We outline briefly how little the results change if one incorporates dividend modeling into the stock-price model. For simplicity we consider a continuous dividend yield D_0, which is the fraction of the stock price received by a stockholder per unit time. The stock-price model then becomes

$$dX_t = (\mu - D_0)X_t \, dt + f(Y_t)X_t \, dW_t,$$

and the volatility process is as before.

When we write the self-financing property for our replicating portfolio, the stream of payment due to dividends should be taken into account; this leads to the coefficient $r - D_0$ in the Black–Scholes equation in the drift (r is unchanged in the discounting term). The corresponding Black–Scholes operator for this model is then

$$\mathcal{L}_{\mathrm{BS}}^D(\sigma) = \frac{\partial}{\partial t} + \frac{1}{2}\sigma^2 x^2 \frac{\partial^2}{\partial x^2} + (r - D_0)x\frac{\partial}{\partial x} - r,$$

which matches our usual definition (1.36) except for the $r - D_0$ replacement in the $\partial/\partial x$ term.

The calculations of Chapter 5 then go through analogously, with the only change being to the operator $\mathcal{L}_2$ defined in (5.13). With the addition of the dividend structure, this becomes

$$\mathcal{L}_2^D = \frac{\partial}{\partial t} + \frac{1}{2}f(y)^2 x^2 \frac{\partial^2}{\partial x^2} + (r - D_0)x\frac{\partial}{\partial x} - r = \mathcal{L}_{\mathrm{BS}}^D(f(y)). \tag{6.4}$$

The first approximation to the price $P^\varepsilon(t, x, y)$ of a European contract is given by $P_0(t, x)$, the Black–Scholes formula for the contract with dividends. It satisfies

$$\langle \mathcal{L}_2^D \rangle P_0 = \mathcal{L}_{\mathrm{BS}}^D(\bar{\sigma}) P_0 = 0,$$

with terminal condition $P_0(T, x) = h(x)$.

The correction for stochastic volatility $\widetilde{P}_1(t, x)$ is given by exactly the same formula (5.41),

$$\widetilde{P}_1(t, x) = -(T - t)\left(V_2 x^2 \frac{\partial^2 P_0}{\partial x^2} + V_3 x^3 \frac{\partial^3 P_0}{\partial x^3}\right),$$

where the effect of the dividend is in P_0. This is because

$$\mathcal{L}_2^D - \langle \mathcal{L}_2^D \rangle = \mathcal{L}_2 - \langle \mathcal{L}_2 \rangle$$

and so the correcting operator in (5.44) is identical (it does not depend on D_0). It is given by (5.45).

For a call option used for calibration through the implied volatility surface, the first approximation is given by the Black–Scholes formula with dividends,

$$P_0(t, x) = xe^{-D_0(T-t)}N(d_{10}) - Ke^{-r(T-t)}N(d_{20}),$$

where

$$d_{10} = \frac{\log(x/K) + (r - D_0 + \frac{1}{2}\bar{\sigma}^2)(T - t)}{\bar{\sigma}\sqrt{T - t}},$$

$$d_{20} = d_{10} - \bar{\sigma}\sqrt{T - t},$$

analogous to (1.37). The correction $\widetilde{P}_1(t, x)$ is given by

$$\widetilde{P}_1(t, x) = \frac{xe^{-D_0(T-t)-d_{10}^2/2}}{\bar{\sigma}\sqrt{2\pi}}\left(V_3\frac{d_{10}}{\bar{\sigma}} + (V_3 - V_2)\sqrt{T - t}\right),$$

replacing (5.54).

The implied volatility remains an affine function of the LMMR, and having estimated the slope a and intercept b, the calibration formulas (5.56)–(5.57) become

$$V_2 = \bar{\sigma}((\bar{\sigma} - b) - a(r - D_0 + \tfrac{3}{2}\bar{\sigma}^2)),$$

$$V_3 = -a\bar{\sigma}^3.$$

In other words, one uses the risk-free rate minus the dividend yield $r - D_0$ to calibrate; then the asymptotic theory is the same with the dividend-adjusted Black–Scholes prices and corresponding Greeks.

6.4 The Second Correction

A natural question is: Why not proceed with the asymptotic expansion described in Chapter 5, obtaining more accurate approximations by calculating higher-order terms? The answer is twofold: the asymptotics is used not only as a computational convenience but also to highlight the dependence on the volatility model parameters. As we include more and more terms, the approximation becomes more accurate but relies on more details or specifics of a model. For example, going one more term to P_2 of the expansion (5.15) introduces three more group parameters that have to be estimated in addition to V_2 and V_3 from the observed skew. This might be acceptable if one wants to capture the geometry of the implied volatility surface more closely, but the real price to pay is that P_2 depends on the present level y of the volatility driving process (Y_t), which is not directly observed; this follows from equation (5.29). In principle, y could be filtered from the stock price returns data, but this would involve *choosing a specific volatility model* $f(\cdot)$. Therefore, we do not pursue this here.

However, it is of interest to pursue the calculation of P_2 for a call or put option to see what correction to the shape of the implied volatility skew arises. That is, we may compute I_2 in the implied volatility expansion

$$I = \bar{\sigma} + \sqrt{\varepsilon}I_1 + \varepsilon I_2 + \cdots,$$

where we know from Section 5.3 that I_1 is affine in the LMMR variable. Of course, I_2 depends on y.

Up to a terminal-layer analysis, the correction to the skew εI_2 is a *quartic in log-moneyness* $\log(K/x)$. It has an extremely complicated dependence on time to maturity. In other words, the second correction captures the out-of-the-money turns of the "smile curve" (implied volatility as a function of moneyness with fixed time to maturity) observed in numerical computations for specific models (see e.g. Figure 2.4).

Notes

This study of the time stability of the estimated group parameters from the S&P 500 implied volatility surface appears in Fouque, Papanicolaou, and Sircar (2000b). Details about modeling dividend payments can be found, for example, in Wilmott et al. (1996) or Duffie (1996).

The implied volatility surface described by the second correction I_2 is work in progress with Yann Samuelides.

7   Hedging Strategies

In this chapter, we look at how the asymptotic analysis helps with the risk management problem of hedging a derivative position. As discussed at the end of Section 2.5, financial institutions often want to eliminate or reduce their exposure to a contingent claim written on an asset by trading in the underlying asset. In an incomplete market, a perfect hedge is not possible and the goal is to find an acceptable trade-off between the risk of a failed hedge and the cost of implementing the hedge. The statistical performance of a strategy is measured by the investor's subjective probability $I\!P$.

7.1   Black–Scholes Delta Hedging

From Section 1.3.2 we know that in a constant volatility environment $\bar{\sigma}$, where the risky asset price $\bar{X}_t$ is a geometric Brownian motion with growth rate μ under $I\!P$, a short position in a European derivative that pays $h(\bar{X}_T)$ can be perfectly hedged by managing the self-financing portfolio made of (at time t) the Delta

$$\frac{\partial P_0}{\partial x}(t, \bar{X}_t)$$

units of the risky asset and

$$e^{-rt}\left(P_0(t, \bar{X}_t) - \bar{X}_t \frac{\partial P_0}{\partial x}(t, \bar{X}_t)\right)$$

units of the riskless asset. Recall that this is simply because the value of such a portfolio at time t is $P_0(t, \bar{X}_t)$, which is precisely $h(\bar{X}_T)$ at maturity; its variation $d(P_0(t, \bar{X}_t))$ is exactly the variation due to the market,

$$\frac{\partial P_0}{\partial x}(t, \bar{X}_t)\, d\bar{X}_t + r\left(P_0(t, \bar{X}_t) - \bar{X}_t \frac{\partial P_0}{\partial x}(t, \bar{X}_t)\right) dt,$$

by Itô's formula and the Black–Scholes equation (5.28) that P_0 satisfies.

115

7.1.1   The Strategy and Its Cost

Let us suppose that we are following the same strategy in the stochastic volatility environment, with $\bar{X}_t$ being replaced by the price process (X_t) associated with
the volatility driving process (Y_t), which is modeled under the subjective measure
$I\!P$ by

$$dX_t = \mu X_t \, dt + f(Y_t) X_t \, dW_t, \tag{7.1}$$

$$dY_t = \alpha(m - Y_t) \, dt + \beta \, d\hat{Z}_t, \tag{7.2}$$

as introduced in Section 2.4.

Such a strategy would duplicate the derivative at maturity because $P_0(T, X_T) =
h(X_T)$, but it would not be self-financing. The portfolio has

$$a_t = \frac{\partial P_0}{\partial x}(t, X_t) \tag{7.3}$$

stocks and

$$b_t = e^{-rt}\left(P_0(t, X_t) - X_t \frac{\partial P_0}{\partial x}(t, X_t) \right)$$

bonds at time t. Its value is

$$a_t X_t + b_t e^{rt} = P_0(t, X_t).$$

Using Itô's formula on $P_0(t, X_t)$, its infinitesimal change is given by

$$dP_0(t, X_t) = \left(\frac{\partial P_0}{\partial t}(t, X_t) + \frac{1}{2} f(Y_t)^2 X_t^2 \frac{\partial^2 P_0}{\partial x^2}(t, X_t) \right) dt + a_t \, dX_t,$$

where we have used (7.3). The infinitesimal change due to the market (the self-
financing part) is given by

$$a_t \, dX_t + r b_t e^{rt} \, dt.$$

Consequently, the infinitesimal cost (positive or negative) of the strategy is given
by the difference

$$dP_0(t, X_t) - a_t \, dX_t - r b_t e^{rt} \, dt = \frac{1}{2}(f(Y_t)^2 - \bar{\sigma}^2) X_t^2 \frac{\partial^2 P_0}{\partial x^2}(t, X_t) \, dt,$$

where we have used the Black–Scholes equation that is satisfied by P_0.

The corresponding **cumulative cost** up to time t is

$$E_0(t) = \frac{1}{2} \int_0^t (f(Y_s)^2 - \bar{\sigma}^2) X_s^2 \frac{\partial^2 P_0}{\partial x^2}(s, X_s) \, ds, \tag{7.4}$$

and the **total cost** is

$$E_0(T) = \frac{1}{2} \int_0^T (f(Y_t)^2 - \bar{\sigma}^2) X_t^2 \frac{\partial^2 P_0}{\partial x^2}(t, X_t) \, dt \qquad (7.5)$$

in *addition* to the initial cost $P_0(0, X_0)$. In other words, the strategy must be further financed or money taken out (consumed) according to the behavior of this integral.

7.1.2 *Averaging Effect*

As the process (Y_t) is running on the fast time scale characterized by large α, the integral

$$\frac{1}{2} \int_0^t f(Y_s)^2 X_s^2 \frac{\partial^2 P_0}{\partial x^2}(s, X_s) \, ds$$

will be close to

$$\frac{1}{2} \langle f^2 \rangle \int_0^t X_s^2 \frac{\partial^2 P_0}{\partial x^2}(s, X_s) \, ds = \frac{1}{2} \bar{\sigma}^2 \int_0^t X_s^2 \frac{\partial^2 P_0}{\partial x^2}(s, X_s) \, ds,$$

because of the rapid oscillations of (Y_t). This is a generalization of the ergodic theorem (3.20) discussed in Chapter 3; it is known as the *averaging* effect. Consequently, the cost E_T – which is the difference between these two integrals – will be small. This is the effect of *centering* $f(Y_t)^2$ with respect to the invariant distribution of Y.

In the spirit of the central limit theorem, we can look at the size of the fluctuations of this small cost. We shall demonstrate that the cost can be written as

$$E_0(t) = \frac{1}{\sqrt{\alpha}} (B_t + M_t) + \mathcal{O}\left(\frac{1}{\alpha}\right),$$

where B_t is a systematic bias and (M_t) a mean-zero martingale, both of which will be computed explicitly.

In order to make this argument rigorous and to evaluate the cost (7.4), one can use the definition (5.30) of ϕ to write

$$f(Y_s)^2 - \bar{\sigma}^2 = (\mathcal{L}_0 \phi)(Y_s),$$

where $\mathcal{L}_0$ is the infinitesimal generator of Y. Using Itô's formula and (7.2), we have

$$d(\phi(Y_s)) = \alpha (\mathcal{L}_0 \phi)(Y_s) \, ds + v \sqrt{2\alpha} \phi'(Y_s) \, d\hat{Z}_s,$$

which leads to

$$(\mathcal{L}_0 \phi)(Y_s) \, ds = \frac{1}{\alpha} \{ d(\phi(Y_s)) - v \sqrt{2\alpha} \phi'(Y_s) \, d\hat{Z}_s \}.$$

The cumulative cost (7.4) is then given by

$$E_0(t) = \frac{1}{2\alpha} \int_0^t X_s^2 \frac{\partial^2 P_0}{\partial x^2}(s, X_s)\{d(\phi(Y_s)) - v\sqrt{2\alpha}\phi'(Y_s)\,d\hat{Z}_s\}.$$

The first integral is computed by using the integration-by-parts formula (1.19) as follows:

$$d\left(X_s^2 \frac{\partial^2 P_0}{\partial x^2}(s, X_s)\phi(Y_s)\right) = X_s^2 \frac{\partial^2 P_0}{\partial x^2}(s, X_s)\,d(\phi(Y_s))$$

$$+ \phi(Y_s)\,d\left(X_s^2 \frac{\partial^2 P_0}{\partial x^2}(s, X_s)\right)$$

$$+ d\left\langle X^2 \frac{\partial^2 P_0}{\partial x^2}, \phi(Y)\right\rangle_s,$$

where the covariation term is given by

$$d\left\langle X^2 \frac{\partial^2 P_0}{\partial x^2}, \phi(Y)\right\rangle_s$$

$$= v\rho\sqrt{2\alpha}\phi'(Y_s)\left(X_s^2 \frac{\partial^3 P_0}{\partial x^3}(s, X_s) + 2X_s \frac{\partial^2 P_0}{\partial x^2}(s, X_s)\right)f(Y_s)X_s\,ds.$$

To simplify the notation, we drop the argument (s, X_s) in the partial derivatives of P_0. We then obtain the cumulative cost

$$E_0(t) = \frac{1}{2\alpha}\left\{X_t^2 \frac{\partial^2 P_0}{\partial x^2}(t, X_t)\phi(Y_t) - X_0^2 \frac{\partial^2 P_0}{\partial x^2}(0, X_0)\phi(Y_0)\right.$$

$$\left. - \int_0^t \phi(Y_s)\,d\left(X_s^2 \frac{\partial^2 P_0}{\partial x^2}\right)\right\}$$

$$- \frac{\rho v}{\sqrt{2\alpha}}\int_0^t f(Y_s)\phi'(Y_s)\left(2X_s^2 \frac{\partial^2 P_0}{\partial x^2} + X_s^3 \frac{\partial^3 P_0}{\partial x^3}\right)ds$$

$$- \frac{v}{\sqrt{2\alpha}}\int_0^t X_s^2 \frac{\partial^2 P_0}{\partial x^2}\phi'(Y_s)\,d\hat{Z}_s. \tag{7.6}$$

The terms within braces are of order $1/\alpha$ since Y is no longer involved in the differentials and since $x^2(\partial^2 P_0/\partial x^2)$ and its derivatives are bounded. The averaging effect will also take place in the second integral, but now the function $f\phi'$ has no reason to be centered with respect to the invariant distribution of Y; hence we know that this term is of order $1/\sqrt{\alpha}$. The same argument applies to the last martingale term. The function ϕ' is not centered with respect to the invariant distribution of Y, but nevertheless it is a zero-mean martingale. In summary, with this hedging strategy we must pay a cost of the form

$$E_0(t) = \frac{1}{\sqrt{\alpha}}(B_t + M_t) + \mathcal{O}\left(\frac{1}{\alpha}\right), \tag{7.7}$$

where

$$B_t = -\frac{\rho v}{\sqrt{2}} \int_0^t f(Y_s) \phi'(Y_s) \left(2X_s^2 \frac{\partial^2 P_0}{\partial x^2} + X_s^3 \frac{\partial^3 P_0}{\partial x^3}\right) ds$$

$$M_t = -\frac{v}{\sqrt{2}} \int_0^t X_s^2 \frac{\partial^2 P_0}{\partial x^2} \phi'(Y_s) \, d\hat{Z}_s.$$

This identifies the drift or bias (B_t) and the mean-zero martingale part of the cost (M_t). Note that the expected bias $I\!E\{B_t\}$ can be of either sign and that the uncorrected Black–Scholes strategy is not *mean self-financing* to the order $1/\sqrt{\alpha}$, which would be the case if the only remaining term of that order were the martingale. In the next section, we introduce a correction that achieves exactly this.

7.2   Mean Self-Financing Hedging Strategy

We propose using the asymptotic method to *remove* the bias in the cumulative cost that was highlighted in the previous calculation. More precisely, the bias B_T will be pushed to the next order by adding a small correction to the Black–Scholes Delta, which has the effect of centering the biasing term. The important feature is that the correction is not simply the Delta of $P_0 + \widetilde{P}_1$ (the asymptotic correction to the price), because that incorporates the market price of volatility risk, which is not relevant to the hedging problem. Here, the controlling probability measure is $I\!P$. The **hedging correction** will depend on the market parameter V_3 discussed in Section 5.2.8, which can be estimated from the implied volatility surface as described in Section 5.3. The analysis separates the effect of the correlation and the effect of the risk premium so that some information from the smile, which is computed in the risk-neutral world, is useful for real-world risk management problems.

We shall correct the hedging strategy as follows. Manage the portfolio made of

$$a_t = \frac{\partial(P_0 + \widetilde{Q}_1)}{\partial x}(t, X_t)$$

shares of the risky asset and

$$b_t = e^{-rt}\left(P_0(t, X_t) + \widetilde{Q}_1(t, X_t) - X_t \frac{\partial(P_0 + \widetilde{Q}_1)}{\partial x}(t, X_t)\right)$$

of the riskless asset, where $\widetilde{Q}_1(t, x)$ solves the partial differential equation

$$\mathcal{L}_{BS}(\bar{\sigma})\widetilde{Q}_1 = V_3\left(2x^2 \frac{\partial^2 P_0}{\partial x^2} + x^3 \frac{\partial^3 P_0}{\partial x^3}\right),$$

with terminal condition

$$\widetilde{Q}_1(T, x) = 0.$$

As we know from (5.42), $\widetilde{Q}_1$ is given explicitly by

$$\widetilde{Q}_1(t, x) = -(T - t)V_3\left(2x^2\frac{\partial^2 P_0}{\partial x^2} + x^3\frac{\partial^3 P_0}{\partial x^3}\right). \tag{7.8}$$

Notice that $\widetilde{Q}_1(t, x)$ is small (of order $1/\sqrt{\alpha}$) because of the order of V_3. In fact, (7.8) is very similar to the formula (5.43) for the pricing correction $\widetilde{P}_1(t, x)$; it is the same except that the part containing Λ is removed, the reason being that the bias in (7.6) does not contain any Λ term to be centered.

The hedging ratio a_t is now given by

$$a_t = \frac{\partial P_0}{\partial x} - \frac{V_3(T - t)}{x}\left(4x^2\frac{\partial^2 P_0}{\partial x^2} + 5x^3\frac{\partial^3 P_0}{\partial x^3} + x^4\frac{\partial^4 P_0}{\partial x^4}\right). \tag{7.9}$$

This is the usual Black–Scholes Delta corrected by a combination involving

$$\text{Gamma} = \frac{\partial^2 P_0}{\partial x^2} \quad \text{and} \quad \text{Epsilon} = \frac{\partial^3 P_0}{\partial x^3}$$

(introduced in Section 5.3) and **Kappa**, defined here as

$$\text{Kappa} = \frac{\partial^4 P_0}{\partial x^4}.$$

With these new hedging ratios, the infinitesimal cost of this hedging strategy is given by

$$dE_1^Q(t) = d(P_0(t, X_t) + \widetilde{Q}_1(t, X_t)) - a_t\, dX_t - rb_t e^{rt}\, dt.$$

Repeating the calculation of the previous section with the new hedging strategy, the *total cost* is now given by

$$E_1^Q(T) = \frac{1}{2}\int_0^T (f(Y_t)^2 - \bar{\sigma}^2)X_t^2\frac{\partial^2 P_0}{\partial x^2}\, dt + \frac{1}{2}\int_0^T (f(Y_t)^2 - \bar{\sigma}^2)X_t^2\frac{\partial^2 \widetilde{Q}_1}{\partial x^2}\, dt$$

$$+ \int_0^T V_3\left(2X_t^2\frac{\partial^2 P_0}{\partial x^2} + X_t^3\frac{\partial^3 P_0}{\partial x^3}\right)dt. \tag{7.10}$$

The second term is of order $1/\alpha$ because (a) the integral is of order $1/\sqrt{\alpha}$, by the averaging effect presented in the previous section, and (b) $\widetilde{Q}_1$ and its derivatives are of order $1/\sqrt{\alpha}$, as seen from (7.8).

The first term is precisely the cumulative cost (7.4) computed in the previous section and obtained in (7.6). By design, the second term in (7.6) combines with the last term in (7.10), owing to the stream of payment generated by the correction, and gives

$$\frac{\rho v}{\sqrt{2\alpha}} \int_0^T \left[2X_t^2 \frac{\partial^2 P_0}{\partial x^2} + X_t^3 \frac{\partial^3 P_0}{\partial x^3} \right] (\langle f\phi' \rangle - f\phi(Y_t))\, dt \qquad (7.11)$$

where we have used the definition (5.40) of V_3. Using the averaging argument of the previous section to handle the integral (7.4), one can show that the integral in (7.11) is of order $1/\sqrt{\alpha}$; hence (7.11) is of order $1/\alpha$. Indeed, the argument requires that we introduce a function $\phi_2(y)$, a solution of the Poisson equation

$$\mathcal{L}_0 \phi_2 = f\phi' - \langle f\phi' \rangle.$$

In the end, all the terms in the corrected cumulative cost (7.10) are of order $1/\alpha$ except the remaining martingale term from (7.6),

$$\frac{1}{\sqrt{\alpha}} M_T = -\frac{v}{\sqrt{2\alpha}} \int_0^T X_s^2 \frac{\partial^2 P_0}{\partial x^2} \phi'(Y_s)\, d\hat{Z}_s, \qquad (7.12)$$

which is of order $1/\sqrt{\alpha}$ but has zero mean. We have thereby removed the systematic bias in (7.7) and so have *reduced the variance* of this small, "nonhedgable" part of the risk due to stochastic volatility. The strategy is now mean self-financing to order $1/\alpha$.

It is important to note that implementing this strategy requires only that we know the parameters $\bar{\sigma}$ and V_3. The calibration of the last parameter comes from the observed implied volatility surface, as discussed in Section 5.3.

We ran 5,000 paths of the expOU model with $\alpha = 200$ and compared the Black–Scholes hedging strategy and the corrected hedge (7.9) for a six-month European call with 200 re-hedgings over that period. The mean of the empirical cumulative cost $E_0(T)$ for the Black–Scholes hedge was \$0.161 or 1.1% of the initial outlay $P_0(0, X_0)$. For the corrected hedge, the mean of the cumulative cost $E_1^Q(T)$ was \$0.122 or 0.8% of the initial cost $P_0(0, X_0) + \widetilde{Q}_1(0, X_0)$. The average saving of almost \$0.04 per option contract hedged outweighs the extra initial cost, which here is \$0.03. Of course, in practice we deal with tens of thousands of option contracts, so these small numbers become very significant!

7.3   Staying Close to the Price

In the previous section we set our criterion for measuring hedging performance in terms of replication at the final time T, as quantified by the total cost. The difference between the price of the derivative and the value of the portfolio is of the order of $\widetilde{P}_1 - \widetilde{Q}_1$, that is, of order $1/\sqrt{\alpha}$. An alternative is to require that the value of the hedging portfolio stay close to the model-predicted derivative price $P(t, X_t, Y_t)$ that satisfies the partial differential equation (5.9)–(5.10). This might be a preferable criterion when we are dealing with American options, for example, and the contract may be exercised against us at any time.

This requirement suggests that we hedge holding the Delta of the corrected price $P_0 + \widetilde{P}_1$ from the pricing analysis of Section 5.2. Consider a portfolio made of

$$a_t = \frac{\partial(P_0 + \widetilde{P}_1)}{\partial x}(t, X_t)$$

shares of the risky asset and

$$b_t = e^{-rt}\left(P_0(t, X_t) + \widetilde{P}_1(t, X_t) - X_t \frac{\partial(P_0 + \widetilde{P}_1)}{\partial x}(t, X_t)\right)$$

of the riskless asset, where $\widetilde{P}_1(t, x) = -(T - t)H(t, x)$ is the price correction of Section 5.2.4 given by (5.43).

For this strategy, the hedging ratio is given by

$$a_t = \frac{\partial P_0}{\partial x} - \frac{(T - t)}{x}\left(2V_2 x^2 \frac{\partial^2 P_0}{\partial x^2} + (V_2 + 3V_3)x^3 \frac{\partial^3 P_0}{\partial x^3} + V_3 x^4 \frac{\partial^4 P_0}{\partial x^4}\right). \quad (7.13)$$

This is again the Black–Scholes Delta corrected by a combination involving the Greeks (Gamma, Epsilon, and Kappa).

We know from Section 5.4 that

$$|P - (P_0 + \widetilde{P}_1)| \le \text{constant} \times 1/\alpha.$$

In other words, the value of the hedging portfolio stays close to the price of the derivative. If we were to use this portfolio and repeat the calculation of Section 7.2 (where we expanded the cost), then the analog of (7.10) would be

$$E_1^P(T) = \frac{1}{2} \int_0^T (f(Y_t)^2 - \bar{\sigma}^2)X_t^2 \frac{\partial^2 P_0}{\partial x^2}\, dt + \frac{1}{2} \int_0^T (f(Y_t)^2 - \bar{\sigma}^2)X_t^2 \frac{\partial^2 \widetilde{P}_1}{\partial x^2}\, dt$$

$$+ \int_0^T H(t, X_t)\, dt, \qquad (7.14)$$

where $H(t, x)$ is the source term depending on V_2 and V_3 given in (5.38). Now however, we are left with an additional term of order $1/\sqrt{\alpha}$ in the cost. The analog of (7.12) becomes

$$\int_0^T (V_2 - 2V_3)X_t^2 \frac{\partial^2 P_0}{\partial x^2}(t, X_t)\, dt - \frac{\nu}{\sqrt{2\alpha}} \int_0^T X_s^2 \frac{\partial^2 P_0}{\partial x^2}\phi'(Y_s)\, d\hat{Z}_s.$$

Notice that, from (5.39) and (5.40),

$$V_2 - 2V_3 = \frac{-\nu\langle\Lambda\phi'\rangle}{\sqrt{2\alpha}}.$$

As a result, the term added to the martingale reflects the premium that the market attaches to volatility risk. This term is of order $1/\sqrt{\alpha}$ and increases the variance of the cumulated cost, but the hedging portfolio replicates the derivative price to order $1/\alpha$ at all times.

Notes

Results about hedging in incomplete markets can be found in Karatzas and Shreve (1998) and El Karoui and Quenez (1995) and references cited there. Variance minimizing strategies and mean self-financing portfolios are described in the survey article by Schweizer (1999), which has references on the history. Hedging strategies for uncorrelated stochastic volatility computed using asymptotic analysis are described in Sircar and Papanicolaou (1999) and Sircar (1999).

8 Application to Exotic Derivatives

In this chapter we show how the asymptotic theory of Chapter 5 can be applied to pricing a wide variety of other popular (European-style) derivative contracts. Some of the simpler ones lead to explicit formulas whereas others require numerical solution, which is easy to implement. American-style derivatives are treated in the next chapter.

All that is needed in the following examples is the mean historical volatility $\bar{\sigma}$ (5.1) together with the slope and intercept parameters a and b from fitting the observed around-the-money implied volatility skew to a line in LMMR, as in equation (5.55). From a, b, $\bar{\sigma}$ we obtain the values of V_2 and V_3 given by (5.56) and (5.57).

8.1 European Binary Options

As an example of a binary (or digital) option with no early exercise feature, we consider a cash-or-nothing call that pays a fixed amount Q on date T if $X_T > K$ but pays zero if $X_T \leq K$. Its payoff function is

$$h(x) = Q\mathbf{1}_{\{x>K\}}, \tag{8.1}$$

where $\mathbf{1}$ denotes the indicator function.

Applying the fast mean-reverting theory of Chapter 5, we compute the stochastic volatility–corrected Black–Scholes price

$$\widetilde{P}(t, x) = P_0(t, x) + \widetilde{P}_1(t, x).$$

The leading term $P_0(t, x)$ is simply the Black–Scholes price of the contract with constant volatility $\bar{\sigma}$. It is given by

$$P_0(t, x) = Qe^{-r(T-t)}N(d_2), \tag{8.2}$$

$$d_2 = \frac{\log(x/K) + (r - \frac{1}{2}\bar{\sigma}^2)(T - t)}{\bar{\sigma}\sqrt{T - t}}, \tag{8.3}$$

where N is the standard normal cumulative distribution function defined in (1.40). This is easily obtained from the risk-neutral pricing formula (1.55), where $(\bar{X}_t)$ is a lognormal process with growth rate r and volatility $\bar{\sigma}$ under the risk-neutral probability measure $\overline{I\!\!P}$. It can be written for this payoff function as

$$P_0(t, x) = Q e^{-r(T-t)} \overline{I\!\!P}\{\bar{X}_T > K \mid \bar{X}_t = x\},$$

leading to (8.2). It is also the solution of the partial differential equation (1.35) with σ replaced by $\bar{\sigma}$ and the terminal condition $P_0(T, x) = Q\mathbf{1}_{\{x>K\}}$.

The correction $\widetilde{P}_1(t, x)$ satisfies

$$\mathcal{L}_{BS}(\bar{\sigma})\widetilde{P}_1(t, x) = \mathcal{A}P_0(t, x),$$

$$\widetilde{P}_1(T, x) = 0,$$

as in Section 5.2.4, using the notation $\mathcal{L}_{BS}$ defined in (1.36) and where we define the differential operator $\mathcal{A}$ by

$$\mathcal{A} = V_3 x^3 \frac{\partial^3}{\partial x^3} + V_2 x^2 \frac{\partial^2}{\partial x^2}; \tag{8.4}$$

the small market coefficients V_2 and V_3 are the same as discussed in Chapter 5.

We know that the solution $\widetilde{P}_1(t, x)$ is given by

$$\widetilde{P}_1(t, x) = -(T - t)\left(V_2 x^2 \frac{\partial^2 P_0}{\partial x^2} + V_3 x^3 \frac{\partial^3 P_0}{\partial x^3} \right),$$

where $P_0(t, x)$ is the Black–Scholes binary price given by (8.2). Explicitly, the correction is given by

$$\widetilde{P}_1(t, x) = Q \frac{(x/K)e^{-d_1^2/2}}{\bar{\sigma}^2 \sqrt{2\pi}} \left(\frac{V_3}{\bar{\sigma}\sqrt{T-t}}(1 - d_1^2) - (V_3 - V_2) d_1 \right),$$

where d_1 is given by (5.51). We plot in Figure 8.1 the Black–Scholes price P_0 and the corrected price $\widetilde{P} = P_0 + \widetilde{P}_1$ of the binary option as a function of the present stock price six months from maturity. We use the parameter values $a, b, \bar{\sigma}$ estimated from S&P 500 data. We notice that, with these values, the correction to the binary price is negative over the price range considered.

8.2 Barrier Options

We now demonstrate how the asymptotic theory handles path-dependent securities by computing the stochastic volatility correction for a down-and-out barrier call option, introduced in Section 1.2.3. Recall that this contract gives the holder

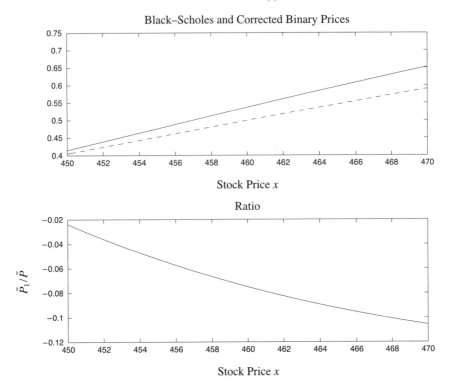

Figure 8.1. The top graph shows Black–Scholes (solid line) and stochastic volatility–corrected (dashed line) binary prices six months from maturity given $K = 460$ and $Q = 1$, with stochastic volatility parameters $\bar{\sigma} = 0.1$, $a = -0.154$, and $b = 0.149$; the interest rate is $r = 0.02$. The bottom graph shows the relative correction.

the right to buy the underlying asset on expiration date T for strike price K unless the asset price has hit the barrier B at any time before T, in which case the contract expires worthless. In what follows, we shall assume that $B < K$.

In this case there is again an explicit solution in the Black–Scholes model, although its derivation becomes more technical. As expected, derivation of a formula for the correction is also technically involved. Some details are given here, but they can be skipped from the point of view of understanding the asymptotic theory.

In the stochastic volatility environment, the price $P(t, x, y)$ of the barrier option satisfies (5.9) with $P(T, x, y) = (x - K)^+$ and boundary condition at $x = B$, $P(t, B, y) = 0$. The asymptotic calculations of Chapter 5 are not affected by the smaller fixed domain $\{x > B\}$ of this problem, as long as we keep track of this boundary condition. Our fast mean-reverting approximation is

$$\widetilde{P}(t, x) = P_0(t, x) + \widetilde{P}_1(t, x),$$

where $P_0(t, x)$ is the Black–Scholes barrier price with constant volatility parameter $\bar{\sigma}$. The stochastic volatility correction $\widetilde{P}_1(t, x)$ satisfies the partial differential equation

$$\mathcal{L}_{BS}(\bar{\sigma})\widetilde{P}_1 = \mathcal{A}P_0 \quad \text{in } x > B \text{ and } t < T, \tag{8.5}$$

with zero terminal and boundary conditions and with $\mathcal{A}$ as in (8.4).

As discussed in Section 1.5.5, the Black–Scholes price $P_0(t, x)$ is obtained by the method of images and is given by

$$P_0(t, x) = C_{BS}(t, x) - \left(\frac{x}{B}\right)^{1-k} C_{BS}\left(t, \frac{B^2}{x}\right),$$

where $C_{BS}(t, x)$ is the Black–Scholes formula for a *call* option, with the volatility parameter $\bar{\sigma}$ and

$$k = 2r/\bar{\sigma}^2.$$

Because of the boundary condition $\widetilde{P}_1(t, B) = 0$, the solution $\widetilde{P}_1$ is not simply given by $-(T - t)\mathcal{A}P_0$ as in the European problems defined on the whole domain $x > 0$. Nevertheless, the following computation (using the method of images) gives an explicit solution.

Computing $\mathcal{A}P_0$, we obtain the right-hand side of (8.5), denoted by $F(t, x)$:

$$
\begin{aligned}
F(t, x) = {} & V_3 x^3 \frac{\partial^3 C_{BS}}{\partial x^3}(t, x) + V_2 x^2 \frac{\partial^2 C_{BS}}{\partial x^2}(t, x) \\
& - \left(\frac{x}{B}\right)^{1-k}\left(V_2 \frac{B^4}{x^2}\frac{\partial^2 C_{BS}}{\partial x^2}\left(t, \frac{B^2}{x}\right)\right. \\
& \left. - V_3 \frac{B^6}{x^3}\frac{\partial^3 C_{BS}}{\partial x^3}\left(t, \frac{B^2}{x}\right) + q\left(t, \frac{B^2}{x}\right)\right), \tag{8.6}
\end{aligned}
$$

with

$$q(t, x) = \kappa_0 C_{BS}(t, x) + \kappa_1 x \frac{\partial C_{BS}}{\partial x}(t, x) + \kappa_2 x^2 \frac{\partial^2 C_{BS}}{\partial x^2}(t, x),$$

$$\kappa_0 = k(k-1)(V_2 - V_3(k+1)), \tag{8.7}$$

$$\kappa_1 = 2kV_2 - 3k\left(\frac{2r}{\bar{\sigma}^2} + 1\right)V_3, \tag{8.8}$$

$$\kappa_2 = -3(k+1)V_3.$$

The method of images consists of defining the mirror operator $\mathcal{M}$ acting on functions $g(t, x)$ by

$$\mathcal{M}g(t, x) = \left(\frac{x}{B}\right)^{1-k} g\left(t, \frac{B^2}{x}\right)$$

and observing that the solution to $\mathcal{L}_{BS}(\bar{\sigma})\widetilde{P}_1 = F(t, x)$ in $x > B$ is given by solving

$$\mathcal{L}_{BS}(\bar{\sigma})v(t, x) = F(t, x) - \mathcal{M}F(t, x)$$

in $x > 0$ and restricting the solution to $x > B$.

From (8.6), we then need only solve

$$\mathcal{L}_{BS}(\bar{\sigma})v(t, x) = 2V_2\left(x^2\frac{\partial^2 C_{BS}}{\partial x^2}(t, x) - \mathcal{M}\left(x^2\frac{\partial^2 C_{BS}}{\partial x^2}(t, x)\right)\right)$$
$$+ q(t, x) - \mathcal{M}q(t, x)$$

on the full domain $x > 0$, $t < T$. Notice that the third-derivative term has disappeared and that the second-derivative term is doubled because of the negative sign induced for every derivative of an image function.

Since the right-hand side is a function minus its image, it can be shown that we can ignore the image terms, solve and then subtract the image of the solution. Thus we need to solve

$$\mathcal{L}_{BS}(\bar{\sigma})u(t, x) = 2V_2 x^2\frac{\partial^2 C_{BS}}{\partial x^2}(t, x) + q(t, x),$$

with zero terminal condition. We can now use the identity (5.42) to write the solution

$$u(t, x) = -(T - t)\left(2V_2 x^2\frac{\partial^2 C_{BS}}{\partial x^2}(t, x) + q(t, x)\right)$$

or, substituting for q,

$$u(t, x) = (2V_2 - 3(k + 1)V_3)x^2\frac{\partial^2 C_{BS}}{\partial x^2}(t, x) + \kappa_1 x\frac{\partial C_{BS}}{\partial x}(t, x) + \kappa_0 C_{BS}(t, x),$$

where κ_0 and κ_1 are as defined in (8.7) and (8.8).

Finally, $v(t, x) = u(t, x) - \mathcal{M}u(t, x)$, and the stochastic volatility correction $\widetilde{P}_1(t, x)$ is the restriction of v to $x > B$:

$$\widetilde{P}_1(t, x) = u(t, x) - \left(\frac{x}{B}\right)^{1-2r/\bar{\sigma}^2} u\left(t, \frac{B^2}{x}\right).$$

The separate components of the formula are easily computed in closed form. Of course, the procedure generalizes to other types of barrier contracts with different payoff functions.

8.3   Asian Options

To illustrate how to correct the Black–Scholes theory for Asian options under fast
mean-reverting stochastic volatility, we consider an Asian average-strike call option with no early exercise. Recall from Section 1.5.5 that the payoff of this contract
on date T is like that of a call option whose strike price is the average of the stock
price between time $t = 0$ and T.

This contract involves the new process

$$I_t = \int_0^t X_s \, ds,$$

and as shown in Section 1.5.5, it introduces a new spatial variable I in the partial
differential equation for the pricing function. We review the essential steps of the
extension of this to stochastic volatility models and then show how to apply the
fast mean-reversion asymptotics to this new partial differential equation. In fact,
we shall see that the main equation for the stochastic volatility correction is modified only slightly.

We start with the stochastic volatility model for the price process (X_t) and
volatility $(f(Y_t))$ given by (2.7), to which we add

$$dI_t = X_t \, dt \tag{8.9}$$

with $I_0 = 0$, assuming that the starting time of the average specified in the contract
is $t = 0$. Under the risk-neutral probability $I\!P^{\star(\gamma)}$, the process (X_t, Y_t, I_t) remains
a Markov process and is described by equations (2.18) and (2.19). The equation
(8.9) for (I_t) is unchanged under the change of measure. The price $P(t, x, y, I)$
of the contract at time $0 \le t < T$ is given by

$$P(t, x, y, I) = I\!E^{\star(\gamma)}\left\{ e^{-r(T-t)}\left(X_T - \frac{I_T}{T}\right)^+ \mid X_t = x, \, Y_t = y, \, I_t = I\right\}. \tag{8.10}$$

It is also obtained as the solution of the partial differential equation

$$\frac{\partial P}{\partial t} + \frac{1}{2}f(y)^2 x^2 \frac{\partial^2 P}{\partial x^2} + r\left(x\frac{\partial P}{\partial x} - P\right) + \rho\beta x f(y)\frac{\partial^2 P}{\partial x \partial y} + \frac{1}{2}\beta^2 \frac{\partial^2 P}{\partial y^2}$$
$$+ (\alpha(m - y) - \beta\Lambda(y))\frac{\partial P}{\partial y} + x\frac{\partial P}{\partial I} = 0, \tag{8.11}$$

which is exactly equation (2.16) with the extra term $x\partial P/\partial I$ coming from (8.9).
The terminal condition is

$$P(T, x, y, I) = \left(x - \frac{I}{T}\right)^+.$$

Under fast mean reversion, we proceed as in Chapter 5 and write (8.11) as

$$\left(\frac{1}{\varepsilon} \mathcal{L}_0 + \frac{1}{\sqrt{\varepsilon}} \mathcal{L}_1 + \widehat{\mathcal{L}_2} \right) P = 0,$$

where $\mathcal{L}_0$, $\mathcal{L}_1$, and $\mathcal{L}_2$ are defined in (5.11), (5.12), and (5.13), respectively; we also define

$$\widehat{\mathcal{L}_2} = \mathcal{L}_2 + x \frac{\partial}{\partial I}.$$

Now the calculations of Section 5.2 go through exactly, with $\widehat{\mathcal{L}_2}$ replacing $\mathcal{L}_2$. Again the corrected price is given by

$$\widetilde{P}(t, x, I) = P_0(t, x, I) + \widetilde{P}_1(t, x, I),$$

where P_0 solves

$$\langle \widehat{\mathcal{L}_2} \rangle P_0 = 0$$

with terminal condition $P_0(T, x, I) = (x - I/T)^+$. Notice that

$$\langle \widehat{\mathcal{L}_2} \rangle = \langle \mathcal{L}_2 \rangle + x \frac{\partial}{\partial I} = \mathcal{L}_{BS}(\bar{\sigma}) + x \frac{\partial}{\partial I}.$$

In other words, P_0 is the Black–Scholes Asian price with constant volatility $\bar{\sigma}$. This is usually solved numerically.

The correction $\widetilde{P}_1(t, x, I)$ satisfies

$$\langle \widehat{\mathcal{L}_2} \rangle \widetilde{P}_1 = \hat{\mathcal{A}} P_0,$$

where

$$\hat{\mathcal{A}} = \sqrt{\varepsilon} \langle \mathcal{L}_1 \mathcal{L}_0^{-1} (\widehat{\mathcal{L}_2} - \langle \widehat{\mathcal{L}_2} \rangle) \rangle,$$

analogous to (5.44). Since the additive extra term in $\widehat{\mathcal{L}_2}$ does not depend on y, we have

$$\widehat{\mathcal{L}_2} - \langle \widehat{\mathcal{L}_2} \rangle = \mathcal{L}_2 - \langle \mathcal{L}_2 \rangle,$$

which implies that $\hat{\mathcal{A}} = \mathcal{A}$. Therefore,

$$\left(\mathcal{L}_{BS}(\bar{\sigma}) + x \frac{\partial}{\partial I} \right) \widetilde{P}_1(t, x, I) = \mathcal{A} P_0(t, x, I)$$

with zero terminal condition. Notice from (8.4) that the computation of the right-hand side $\mathcal{A} P_0(t, x, I)$ involves only x-derivatives. This equation is solved numerically, too.

Notes

Pricing of the exotic contracts discussed here under the constant volatility Black–Scholes model is solved, explicitly or numerically, in Wilmott et al. (1996) by differential equations methods and in Musiela and Rutkowski (1997) by probabilistic representations.

The correction for the barrier option in Section 8.2 appeared in Fouque, Papanicolaou, and Sircar (1999a), where it was written in terms of the Greeks Vega and Rho ($\partial C_{BS}/\partial r$). The formula given there is identical to the one here using the identities

$$\frac{\partial C_{BS}}{\partial \sigma} = \sigma(T - t)x^2 \frac{\partial^2 C_{BS}}{\partial x^2},$$

$$\frac{\partial C_{BS}}{\partial r} = -(T - t)\left(C_{BS} - x\frac{\partial C_{BS}}{\partial x}\right).$$

The explicit correction for the binary option in Section 8.1 and the correction for the Asian contract in Section 8.3 appear for the first time in this chapter.

9 Application to American Derivatives

We now look at American-style derivative contracts, introduced in Section 1.2.2. Recall that they give the holder the right of early exercise and so the date that the contract is terminated is not known beforehand, unlike in the European case. As we reviewed in Section 1.5.4, for an American put option under the Black–Scholes model, the *no-arbitrage* pricing function satisfies a free boundary value problem characterized by the system of equations (1.73) with boundary conditions (1.74)–(1.77). This is a much harder problem than the European pricing problem, and there are no explicit solutions in general; it must be solved numerically. Nevertheless, we show in this chapter that the asymptotic method for correcting the Black–Scholes price for stochastic volatility can be extended to contracts with the early exercise feature, simplifying considerably the two-dimensional free boundary problems that arise in these models. Furthermore, the correction depends only on the universal group parameters V_2 and V_3 that we estimate from the implied volatility surface, as discussed in Chapter 6.

We concentrate on the American put option, which is the best-known American-style derivative. The procedure applies to the American version of any of the other derivatives we considered in Chapter 8.

9.1 American Problem under Stochastic Volatility

We assume the stochastic volatility model (2.7) and that the market selects a unique pricing measure $I\!P^{\star(\gamma)}$ that is reflected in liquidly traded around-the-money European option prices. Prices of other derivative securities must be priced with respect to this measure if there are to be no arbitrage opportunities. Therefore, the American put price $P(t, x, y)$ is given by

$$P(t, x, y) = \sup_{t \leq \tau \leq T} I\!E^{\star(\gamma)}\{e^{-r(\tau-t)}(K - X_\tau)^+ \mid X_t = x, Y_t = y\},$$

where the supremum is taken over all stopping times $\tau \in [t, T]$ and the process (X_t, Y_t) satisfies (2.18)–(2.19) under $I\!P^{\star(y)}$.

This function $P(t, x, y)$ again satisfies a free boundary problem analogous to (1.73), with the additional spatial variable y, and the free boundary is now a surface that can be written $x = x_{\mathrm{fb}}(t, y)$ and must be determined as part of the problem:

$$P(t, x, y) = K - x \quad \text{for } x < x_{\mathrm{fb}}(t, y), \tag{9.1}$$

$$\frac{\partial P}{\partial t} + \frac{1}{2} f(y)^2 x^2 \frac{\partial^2 P}{\partial x^2} + \rho \beta x f(y) \frac{\partial^2 P}{\partial x \partial y} + \frac{1}{2} \beta^2 \frac{\partial^2 P}{\partial y^2} + r \left(x \frac{\partial P}{\partial x} - P \right)$$

$$+ (\alpha(m - y) - \beta \Lambda(y)) \frac{\partial P}{\partial y} = 0 \quad \text{for } x > x_{\mathrm{fb}}(t, y), \tag{9.2}$$

with

$$P(T, x, y) = (K - x)^+, \tag{9.3}$$

$$x_{\mathrm{fb}}(T, y) = K.$$

Also, P, $\frac{\partial P}{\partial x}$, and $\frac{\partial P}{\partial y}$ are continuous across the boundary $x_{\mathrm{fb}}(t, y)$, so that

$$P(t, x_{\mathrm{fb}}(t, y), y) = (K - x_{\mathrm{fb}}(t, y))^+, \tag{9.4}$$

$$\frac{\partial P}{\partial x}(t, x_{\mathrm{fb}}(t, y), y) = -1,$$

$$\frac{\partial P}{\partial y}(t, x_{\mathrm{fb}}(t, y), y) = 0.$$

The full problem is shown in Figure 9.1.

9.2   Stochastic Volatility Correction for an American Put

We use the notation of Chapter 5:

$$\varepsilon = \frac{1}{\alpha},$$

$$\beta = \frac{\sqrt{2}\nu}{\sqrt{\varepsilon}},$$

$$P^\varepsilon(t, x, y) = P(t, x, y),$$

$$\mathcal{L}^\varepsilon = \frac{1}{\varepsilon} \mathcal{L}_0 + \frac{1}{\sqrt{\varepsilon}} \mathcal{L}_1 + \mathcal{L}_2,$$

where $\mathcal{L}_0$ is the infinitesimal generator of the mean-reverting OU process (scaled by $1/\alpha$), $\mathcal{L}_1$ contains the mixed derivative (from the correlation) and the market

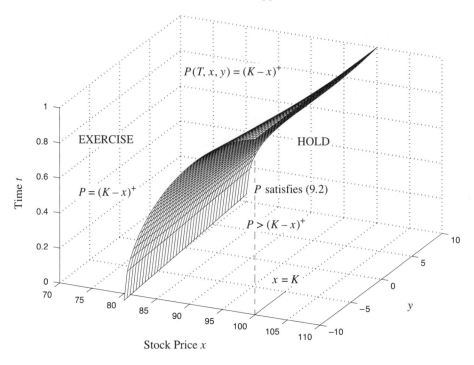

Figure 9.1. The full problem for the American put under stochastic volatility. The free boundary conditions on the surface are given by (9.4).

price of risk γ, and $\mathcal{L}_2$ is $\mathcal{L}_{BS}(f(y))$. These are defined in (5.11)–(5.13). The main equation (9.2) can be written as

$$\mathcal{L}^\varepsilon P^\varepsilon(t, x, y) = 0 \quad \text{in } x > x_{fb}^\varepsilon(t, y),$$

where we denote the free boundary surface by $x_{fb}^\varepsilon(t, y)$ to stress the dependence on ε. Note that the exercise boundary is (in general) a surface $F^\varepsilon(t, x, y) = 0$, which we write as $x = x_{fb}^\varepsilon(t, y)$.

9.2.1 Expansions

As in Chapter 5, we look for an asymptotic solution of the form

$$P^\varepsilon(t, x, y) = P_0(t, x, y) + \sqrt{\varepsilon}\, P_1(t, x, y) + \varepsilon P_2(t, x, y) + \cdots, \quad (9.5)$$

$$x_{fb}^\varepsilon(t, y) = x_0(t, y) + \sqrt{\varepsilon} x_1(t, y) + \varepsilon x_2(t, y) + \cdots, \quad (9.6)$$

which converges as $\varepsilon \downarrow 0$. We have expanded the formula for the free boundary surface as well.

Our strategy for constructing a solution will be to expand the equations and boundary conditions in powers of ε, substituting the expansions (9.5) and (9.6). We then look at the equations at each order (in both the hold and exercise regions) and take the dividing boundary for each subproblem to be $x_0(t, y)$, which is accurate to principal order. Thus, the extension or truncation of the hold region to the x_0 boundary is assumed to introduce only an $\mathcal{O}(\sqrt{\varepsilon})$ error into each term $P_j(t, x, y)$ of the expansion for the price. This will be true up to a region of width $\mathcal{O}(\sqrt{\varepsilon})$ about x_0. When the stock price is so close to the exercise boundary, we do not expect the asymptotics to be accurate because the contract likely does not exist long enough for the "averaging effects" of fast mean-reverting volatility to take hold. This is exactly as we found in Section 5.5 for a European option close to the expiration date, when the asymptotic approximation is not valid.

The expansion of the partial differential equation $\mathcal{L}^\varepsilon P^\varepsilon = 0$ in the hold region is as in the European case:

$$\frac{1}{\varepsilon}\mathcal{L}_0 P_0 + \frac{1}{\sqrt{\varepsilon}}(\mathcal{L}_0 P_1 + \mathcal{L}_1 P_0) + (\mathcal{L}_0 P_2 + \mathcal{L}_1 P_1 + \mathcal{L}_2 P_0)$$
$$+ \sqrt{\varepsilon}(\mathcal{L}_0 P_3 + \mathcal{L}_1 P_2 + \mathcal{L}_2 P_1) + \cdots = 0. \qquad (9.7)$$

Keeping terms up to $\sqrt{\varepsilon}$, we expand the free boundary conditions (9.4) as follows:

$$P_0(t, x_0(t, y), y) + \sqrt{\varepsilon}\left(x_1(t, y)\frac{\partial P_0}{\partial x}(t, x_0(t, y), y) + P_1(t, x_0(t, y), y)\right)$$
$$= K - x_0(t, y) - \sqrt{\varepsilon}x_1(t, y), \qquad (9.8)$$

$$\frac{\partial P_0}{\partial x}(t, x_0(t, y), y) + \sqrt{\varepsilon}\left(x_1(t, y)\frac{\partial^2 P_0}{\partial x^2}(t, x_0(t, y), y) + \frac{\partial P_1}{\partial x}(t, x_0(t, y), y)\right)$$
$$= -1, \qquad (9.9)$$

$$\frac{\partial P_0}{\partial y}(t, x_0(t, y), y) + \sqrt{\varepsilon}\left(x_1(t, y)\frac{\partial^2 P_0}{\partial x \partial y}(t, x_0(t, y), y) + \frac{\partial P_1}{\partial y}(t, x_0(t, y), y)\right)$$
$$= 0, \qquad (9.10)$$

where the partial derivatives are taken to mean the one-sided derivatives into the region $x > x_0(t, y)$ because we expect – by analogy with the Black–Scholes American pricing problem – that the pricing function will not be smooth across the free boundary. However, it is smooth inside either region.

The terminal condition gives $P_0(T, x, y) = (K - x)^+$ and $P_1(T, x, y) = 0$, and the condition $P^\varepsilon = (K - x)^+$ in the exercise region gives that $P_0(t, x, y) = (K - x)^+$ and $P_1(t, x, y) = 0$ in that region.

9.2.2 First Approximation

To largest order in ε, we have the following problem:

$$\mathcal{L}_0 P_0(t, x, y) = 0 \qquad \text{in } x > x_0(t, y),$$

$$P_0(t, x, y) = (K - x)^+ \quad \text{in } x < x_0(t, y),$$

$$P_0(t, x_0(t, y), y) = (K - x_0(t, y))^+,$$

$$\frac{\partial P_0}{\partial x}(t, x_0(t, y), y) = -1.$$

Since $\mathcal{L}_0$ is the generator of an ergodic Markov process acting on the variable y, the usual argument described at the end of Section 3.2.3 implies that P_0 does not depend on y on each side of x_0. It therefore cannot depend on y on the surface x_0 either, and so $x_0 = x_0(t)$ also does not depend on y.

Recall that $\mathcal{L}_1$ contains y-derivatives in both terms, so that $\mathcal{L}_1 P_0 = 0$, and the next order gives

$$\mathcal{L}_0 P_1(t, x, y) = 0 \quad \text{in } x > x_0(t),$$

$$P_1(t, x, y) = 0 \quad \text{in } x < x_0(t),$$

$$P_1(t, x_0(t), y) = 0,$$

$$x_1(t, y) \frac{\partial^2 P_0}{\partial x^2}(t, x_0(t)) + \frac{\partial P_1}{\partial x}(t, x_0(t), y) = 0.$$

By the same argument, P_1 also does not depend on y: $P_1 = P_1(t, x)$.

From the $\mathcal{O}(1)$ terms in (9.7), we have

$$\mathcal{L}_0 P_2(t, x, y) + \mathcal{L}_2 P_0(t, x) = 0 \quad \text{in } x > x_0(t),$$

$$P_2(t, x, y) = 0 \quad \text{in } x < x_0(t),$$

(9.11)

since $\mathcal{L}_1 P_1 = 0$. In the region $x > x_0(t)$, this is a Poisson equation over $-\infty < y < \infty$, because the exercise boundary does not depend on y to principal order. There is no solution unless $\mathcal{L}_2 P_0$ has mean zero with respect to the invariant measure of the OU process Y_t:

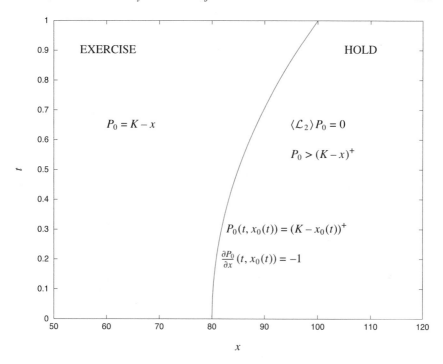

Figure 9.2. The problem for $P_0(t, x)$, which is exactly the Black–Scholes American put pricing problem with average volatility $\bar{\sigma}$.

$$\langle \mathcal{L}_2 P_0 \rangle = 0,$$

where $\langle \cdot \rangle$ is the expectation with respect to the invariant measure (5.23) of the OU process (as introduced in Section 3.2.3). Since $\mathcal{L}_2$ depends on y only through the $f(y)$ coefficient, we have $\langle \mathcal{L}_2 P_0 \rangle = \langle \mathcal{L}_2 \rangle P_0$ and

$$\langle \mathcal{L}_2 \rangle = \mathcal{L}_{\mathrm{BS}}(\bar{\sigma}) = \frac{\partial}{\partial t} + \frac{1}{2}\bar{\sigma}^2 x^2 \frac{\partial^2}{\partial x^2} + r\left(x \frac{\partial}{\partial x} - \cdot\right),$$

where $\bar{\sigma}^2 = \langle f^2 \rangle$. Thus $P_0(t, x)$ and $x_0(t)$ satisfy the problem shown in Figure 9.2, which is exactly the Black–Scholes American put problem with constant volatility $\bar{\sigma}$.

There is no explicit solution for $P_0(t, x)$ or $x_0(t)$, and we discuss how to compute them numerically, along with the stochastic volatility correction, in Section 9.3.

9.2.3   The Stochastic Volatility Correction

We now look for the function $\sqrt{\varepsilon}\,P_1$, which corrects the Black–Scholes American put pricing function P_0 for fast mean-reverting stochastic volatility.

The $\mathcal{O}(\sqrt{\varepsilon})$ terms in (9.7) give that

$$\mathcal{L}_0 P_3(t, x, y) + \mathcal{L}_1 P_2(t, x, y) + \mathcal{L}_2 P_1(t, x) = 0 \quad \text{in } x > x_0(t),$$

$$P_3(t, x, y) = 0 \quad \text{in } x < x_0(t).$$

In the hold region $x > x_0(t)$, this is a Poisson equation for P_3 over $-\infty < y < \infty$. It has no solution unless

$$\langle \mathcal{L}_1 P_2 + \mathcal{L}_2 P_1 \rangle = 0.$$

Substituting for $P_2(t, x, y)$ with

$$P_2 = -\mathcal{L}_0^{-1}(\mathcal{L}_2 - \langle \mathcal{L}_2 \rangle)\, P_0$$

from (9.11), this condition is

$$\langle \mathcal{L}_2 P_1 - \mathcal{L}_1 \mathcal{L}_0^{-1}(\mathcal{L}_2 - \langle \mathcal{L}_2 \rangle) P_0 \rangle = 0,$$

where

$$\langle \mathcal{L}_2 P_1 \rangle = \langle \mathcal{L}_2 \rangle P_1 = \mathcal{L}_{BS}(\bar{\sigma}) P_1,$$

since P_1 does not depend on y.

As is by now our usual procedure, we write the equation for

$$\widetilde{P}_1(t, x) = \sqrt{\varepsilon}\, P_1(t, x),$$

so that ε will be absorbed in with the other parameters. Using the notation

$$\mathcal{A} = \sqrt{\varepsilon}\langle \mathcal{L}_1 \mathcal{L}_0^{-1}(\mathcal{L}_2 - \langle \mathcal{L}_2 \rangle) \rangle,$$

as in Section 5.2.5, the equation determining $\widetilde{P}_1$ in the hold region is

$$\mathcal{L}_{BS}(\bar{\sigma}) \widetilde{P}_1 = \mathcal{A} P_0, \tag{9.12}$$

as P_0 does not depend on y.

The operator $\mathcal{A}$ is computed explicitly in (5.45) as

$$\mathcal{A} = V_3 x^3 \frac{\partial^3}{\partial x^3} + V_2 x^2 \frac{\partial^2}{\partial x^2},$$

where the parameters $V_2 = V_2(a, b, \bar{\sigma})$ and $V_3 = V_3(a, b, \bar{\sigma})$ are obtained from the mean historical volatility and the slope and intercept of the European options

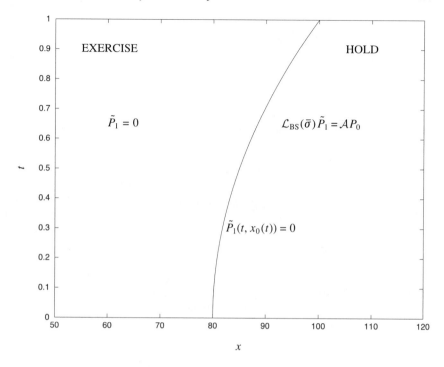

Figure 9.3. The fixed boundary problem for $\widetilde{P}_1(t, x)$.

implied volatility as a linear function of the LMMR (5.55) through the relations (5.56) and (5.57) *exactly as for European derivatives*. Indeed, $\mathcal{A}$ is exactly the operator appearing in the equation for the correction to European securities prices (5.37) and barrier options (8.5).

Thus, in the region $x > x_0(t)$, $\widetilde{P}_1(t, x)$ satisfies

$$\mathcal{L}_{\mathrm{BS}}(\bar{\sigma})\widetilde{P}_1 = V_3 x^3 \frac{\partial^3 P_0}{\partial x^3} + V_2 x^2 \frac{\partial^2 P_0}{\partial x^2}, \qquad (9.13)$$

where $P_0(t, x)$ is the Black–Scholes American put price. It can be shown that $P_0(t, x)$ is bounded with bounded derivatives inside the hold region $x > x_0(t)$, $t < T$, and that the discontinuity of its second x-derivative across $x_0(t)$ is not a difficulty for the $\widetilde{P}_1$ problem (either analytically or numerically).

The complete problem for $\widetilde{P}_1$ is shown in Figure 9.3. This is a *fixed* boundary problem for $\widetilde{P}_1(t, x)$: the boundary $x_0(t)$ is the free boundary determined from the

P_0 problem. However, this boundary is determined up to an error of $\sqrt{\varepsilon}$, so there is an $\mathcal{O}(\sqrt{\varepsilon})$ error in $\widetilde{P}_1$ within an $\mathcal{O}(\sqrt{\varepsilon})$ neighborhood of x_0. The asymptotic approximation is good outside this neighborhood of $x_0(t)$.

Observe that the function

$$-(T - t)\mathcal{A}P_0(t, x)$$

satisfies the equation (9.13) in the hold region $x > x_0(t)$, as in formula (5.43), but it does not satisfy the zero boundary condition required for $\widetilde{P}_1$. Consequently, $\widetilde{P}_1$ is found numerically after obtaining the numerical solution P_0.

The mathematical justification of the formal asymptotic expansion in the American case combines the analysis for the European case presented in Section 5.4 with "penalization" methods for free boundary value problems.

9.2.4 Uncorrelated Volatility

We know from (5.39) and (5.40) that V_3 is proportional to ρ. When volatility shocks are uncorrelated with stock-price shocks, $\rho = 0$ and consequently $V_3 = 0$ and $a = 0$ by (5.57). In this case, in the hold region we have

$$\mathcal{L}_{BS}(\bar{\sigma})(P_0 + \widetilde{P}_1) = V_2 x^2 \frac{\partial^2 P_0}{\partial x^2},$$

so that

$$\mathcal{L}_{BS}\left(\sqrt{\bar{\sigma}^2 - 2V_2}\right)(P_0 + \widetilde{P}_1) = -V_2 x^2 \frac{\partial^2 \widetilde{P}_1}{\partial x^2} = \mathcal{O}(\varepsilon),$$

where we have used $V_2 = \mathcal{O}(\sqrt{\varepsilon})$ and assumed sufficient smoothness in $\widetilde{P}_1$ away from $x_0(t)$ such that $\partial^2 \widetilde{P}_1/\partial x^2 = \mathcal{O}(\sqrt{\varepsilon})$.

It follows that the corrected American price $\widetilde{P} = P_0 + \widetilde{P}_1$ is, up to $\mathcal{O}(\varepsilon)$, the solution of the Black–Scholes American put pricing problem with *corrected effective volatility*

$$\tilde{\sigma} = \sqrt{\bar{\sigma}^2 - 2V_2}.$$

That is, in the absence of correlation, the first-order effect of fast mean-reverting stochastic volatility is simply a volatility level correction; because $V_2 < 0$ whenever at-the-money implied volatility $b > \bar{\sigma}$, this shift is typically upward. This is as we discussed for European securities in Section 5.2.8.

Solving this problem also gives the corrected exercise boundary, which is exactly the Black–Scholes boundary associated with $\tilde{\sigma}$.

9.2.5 *Probabilistic Representation*

From Figure 9.3, $\widetilde{P}_1$ can also be represented as an expectation of a functional of the geometric Brownian motion $\bar{X}_t$ defined by

$$d\bar{X}_t = r\bar{X}_t \, dt + \bar{\sigma}\bar{X}_t \, d\bar{W}_t,$$

where $\bar{W}_t$ is a standard Brownian motion under the probability $\overline{I\!P}$. The process $\bar{X}_t$ is *stopped* at the boundary $x_0(t)$, so that

$$\widetilde{P}_1(t, x) = \overline{I\!E}\left\{-\int_t^T e^{-r(s-t)} \mathcal{A}P_0(s, \bar{X}_s)\mathbf{1}_{\{\bar{X}_u > x_0(u) \text{ for all } t \leq u \leq s\}} \, ds \mid \bar{X}_t = x\right\}.$$

9.3 Numerical Computation

We are interested in computing numerically the stochastic volatility–corrected American put price $P_0(t, x) + \widetilde{P}_1(t, x)$ away from the exercise boundary $x_0(t)$. In order to do this we will use implicit finite differences, first to determine the Black–Scholes price $P_0(t, x)$ satisfying the system (1.72). Then, we find the constant volatility exercise boundary $x_0(t)$ as the largest value of x where $P(t, x)$ coincides with the payoff function $(K - x)^+$.

Finally, we solve the fixed boundary problem (9.13) on the same grid in the region $x > x_0(t)$, $t < T$.

9.3.1 *Solution of the Black–Scholes Problem*

We give references in the notes for the numerical procedure used here to solve the Black–Scholes American pricing problem. It employs the backward Euler finite-difference stencil on a uniform grid (after a change to logarithmic stock-price coordinates). The constraint that the function $P_0(t, x)$ lie above the payoff function is enforced by using the "projected successive over relaxation" (PSOR) algorithm, in which an iterative method is used to solve the implicit time-stepping equations while preserving the constraint between iterations.

Explicit treelike methods are popular in the industry, but we prefer the stability of implicit methods that allow us to take a relatively large time step. In addition we recover the whole pricing function, which allows us to visualise the quality of the solution and the effect of changing parameters. We shall not give the numerical details here; the reader can find them in sources given in the notes at the end of the chapter.

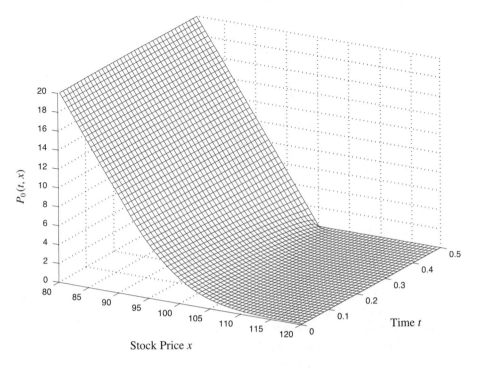

Figure 9.4. Numerical solution for $P_0(t, x)$, the Black–Scholes American put pricing function with $\bar{\sigma} = 0.1$, as estimated from historical S&P 500 data with $K = 100$, $T = 0.5$, and $r = 0.02$, using the backward Euler–PSOR method.

In Figure 9.4 we show a numerical approximation to $P_0(t, x)$, the Black–Scholes American put pricing function.

9.3.2 Computation of the Correction

To compute the correction for stochastic volatility $\widetilde{P}_1(t, x)$ to the Black–Scholes price we have found, we need to solve the fixed boundary problem (9.13) in the region $\{(t, x) : x > x_0(t), 0 \leq t \leq T\}$. Outside this region, the correction is zero. We are most interested in $\widetilde{P}_1$ away from the exercise boundary, particularly around the money $x \approx K$. Close to the curve $x_0(t)$ or the final time T, we do not expect the asymptotic approximation to be valid because the contract will likely expire soon, and volatility will not appear to be fast mean-reverting over such a short time scale (as explained in Section 9.2.1).

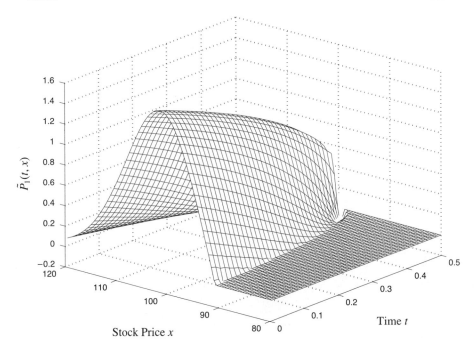

Figure 9.5. The correction $\widetilde{P}_1(t, x)$ to the Black–Scholes American put price to account for fast mean-reverting stochastic volatility, using the parameters estimated from S&P 500 implied volatilies ($a = -0.154$, $b = 0.149$) and from historical index data ($\bar{\sigma} = 0.1$). The correction is computed via backward Euler using the numerical solution shown in Figure 9.4.

In Figure 9.5, we show $\widetilde{P}_1(t, x)$ using parameter values estimated from the 1994 S&P 500 implied volatility surface. The effect of the correction is most around the strike price, and Figure 9.6 shows the Black–Scholes $P_0(0, x)$ and corrected $P_0(0, x) + \widetilde{P}_1(0, x)$.

Notes

Details of the derivation of the free boundary formulation of the American pricing problem are given in Lamberton and Lapeyre (1996). Numerical solution of the Black–Scholes problem is discussed there and in Wilmott et al. (1996), which contains details of the PSOR algorithm. The penalization method for this type of problem can be found in Bensoussan and Lions (1982).

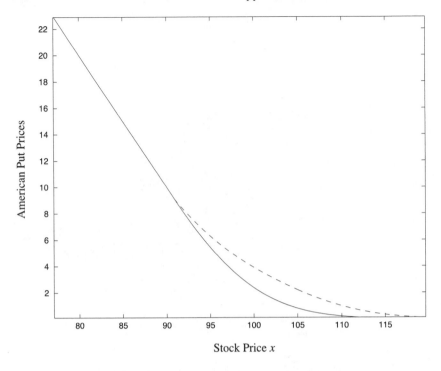

Figure 9.6. Effect of the stochastic volatility correction on American put prices at time $t = 0$. The solid line shows the Black–Scholes American put price $P_0(0, x)$ with the constant historical volatility $\bar{\sigma} = 0.1$, and the dashed line shows the corrected price $P_0(0, x) + \widetilde{P}_1(0, x)$ using the S&P 500 parameters described in the caption of Figure 9.5.

Further details of the asymptotic and numerical calculations presented in this chapter can be found in Fouque, Papanicolaou, and Sircar (1999b).

10 Generalizations

In this chapter, we discuss various generalizations of the asymptotic method that exploits separation of scales or fast mean reversion in the volatility. As we demonstrate, the method is far-reaching and can be implemented in contexts other than pricing and hedging – as, for instance, in the optimal allocation problem of Section 10.1. We also present a martingale approach to the approximation problem which shows that the method summarized in Chapter 6 can also be implemented in non-Markovian situations, where partial differential equations (PDEs) are not available. The application to fixed-income markets is the subject of Chapter 11.

10.1 Portfolio Optimization under Stochastic Volatility

We look at the optimal asset allocation problem facing an investor who must decide how much of his or her wealth to invest in a risky asset (X_t) and how much in the riskless bond in order to maximize expected utility of wealth at the final time T. First, we shall briefly review the case of constant volatility, when the optimal strategy is known as the Merton solution. Then we look at the same problem when the price of the risky asset has randomly changing fast mean-reverting volatility. The asymptotic analysis gives a perturbation of a constant volatility Merton problem, with an effective volatility different from the $\bar{\sigma}$ estimated for pricing problems. However, this solution requires us to track the unobservable volatility, so we present a suboptimal but more practical solution in Section 10.1.3.

10.1.1 Constant Volatility Merton Problem

We take the lognormal model

$$dX_t = \mu X_t \, dt + \sigma X_t \, dW_t$$

145

for the stock price (X_t) and a constant interest rate r. An investor has wealth $\mathcal{W}_t$ at time t made of a_t stocks and b_t bonds,

$$W_t = a_t X_t + b_t e^{rt}.$$

He trades in a self-financing manner so that his wealth process satisfies

$$dW_t = a_t\, dX_t + b_t\, d(e^{rt}).$$

Denoting by u_t and $1 - u_t$ the fraction of wealth in the stock and bond (respectively) at time t, we have $a_t = u_t \mathcal{W}_t/X_t$ and $b_t = (1 - u_t)\mathcal{W}_t/e^{rt}$; hence

$$dW_t = u_t \mathcal{W}_t \frac{dX_t}{X_t} + (1 - u_t)r\mathcal{W}_t\, dt.$$

Using the geometric Brownian structure of the stock price, $(\mathcal{W}_t)$ satisfies the equation

$$dW_t = \mathcal{W}_t[(r + (\mu - r)u_t)\, dt + \sigma u_t\, dW_t],$$

with initial wealth $\mathcal{W}_0 = w$. Note that this is a closed equation because X does not appear, and in particular this implies that $(\mathcal{W}_t)$ is a Markov process by itself if the control u_t is chosen through a Markov control policy (i.e., if u_t is a function of $(t, \mathcal{W}_t)$).

The goal of the investor is to choose the strategy (u_t) to maximize his expected utility at some given finite terminal time T,

$$I\!\!E\{U(\mathcal{W}_T)\},$$

where we shall restrict ourselves to the example of a so-called HARA (hyperbolic absolute risk-averse) utility function

$$U(w) = w^p/p$$

for some $0 < p < 1$. The expectation is with respect to the subjective probability. In the spirit of dynamic programming, we define

$$V(t, w) = \sup_u I\!\!E\left\{\frac{\mathcal{W}_T^p}{p} \;\middle|\; \mathcal{W}_t = w\right\},$$

the maximum expected utility as a function of the starting time t and the starting wealth w.

By the Bellman principle (for which we give references in the notes at the end of the chapter), $V(t, x)$ satisfies the nonlinear Hamilton–Jacobi–Bellman (HJB) partial differential equation

$$\frac{\partial V}{\partial t} + \sup_u\left\{\frac{1}{2}\sigma^2 u^2 w^2 \frac{\partial^2 V}{\partial w^2} + (r + (\mu - r)u)w\frac{\partial V}{\partial w}\right\} = 0$$

with terminal condition $V(T, w) = w^p/p$, where in this Markovian framework the *control* turns out to be $u = u(t, w)$, a function of time and wealth. When the HJB equation has a solution to which Itô's formula can be applied, the verification that it is the optimal expected utility follows readily.

The special form of the chosen utility function motivates the representation

$$V(t, w) = \frac{w^p}{p} v(t).$$

This leads to the linear ordinary differential equation for $v(t)$,

$$\frac{dv}{dt} = pv \sup_u \left\{ \frac{1}{2}\sigma^2 u^2 (p - 1) + (\mu - r)u + r \right\} \tag{10.1}$$

with $v(T) = 1$. The supremum is attained at

$$u^\star = \frac{(\mu - r)}{\sigma^2 (1 - p)}. \tag{10.2}$$

Note that $u^\star$ is simply a constant that is positive when $\mu > r$. This implies that the optimal strategy does not depend on the wealth, which allows *mutual funds* where the portfolio allocation is independent of the wealth of the participant. Moreover, the fraction invested in the stock is constant in time.

The corresponding maximum expected utility is given by

$$V(t, w) = \frac{w^p}{p} \exp\left(\left(r + \frac{(\mu - r)^2}{2\sigma^2 (1 - p)} \right) p(T - t) \right). \tag{10.3}$$

10.1.2 *Stochastic Volatility Merton Problem*

We now tackle the same problem when the market model incorporates random volatility:

$$dX_t = \mu X_t \, dt + f(Y_t) X_t \, dW_t,$$

$$dY_t = \alpha(m - Y_t) \, dt + \beta\left(\rho \, dW_t + \sqrt{1 - \rho^2} \, dZ_t \right).$$

Again, for clarity of exposition, we explicitly write the driving process (Y_t) as an OU process.

Now the wealth process $(\mathcal{W}_t)$ satisfies

$$d\mathcal{W}_t = \mathcal{W}_t[(r + (\mu - r)u_t) \, dt + u_t f(Y_t) \, dW_t],$$

where again u_t denotes the fraction of wealth in the stock. If both wealth and volatility are observable then we are in a Markovian case, where u_t will be of the form $u = u(t, w, y)$. This is an ideal but unrealistic scenario, since we know that in general volatility is not directly observable. The maximum expected HARA utility is now

$$V(t, w, y) = \sup_u I\!E\left\{\frac{W_T^p}{p} \;\middle|\; W_t = w,\, Y_t = y\right\}.$$

The HJB equation becomes

$$\frac{\partial V}{\partial t} + \sup_u \left\{ \frac{1}{2} f(y)^2 u^2 w^2 \frac{\partial^2 V}{\partial w^2} + \beta\rho f(y) uw \frac{\partial^2 V}{\partial w \partial y} \right.$$

$$\left. + (r + (\mu - r)u)w \frac{\partial V}{\partial w} \right\} + \alpha\mathcal{L}_0 V = 0,$$

with terminal condition $V(T, w, y) = w^p/p$. Recall that $\alpha\mathcal{L}_0$ is the infinitesimal generator of the OU process (Y_t) defined in (5.11).

Using the transformation $V(t, w, y) = (w^p/p)v(t, y)$, we obtain the following equation for v:

$$\frac{\partial v}{\partial t} + p\sup_u \left\{ \frac{1}{2}(p-1)f(y)^2 u^2 v \right.$$

$$\left. + u\left[\beta\rho f(y)\frac{\partial v}{\partial y} + (\mu - r)v \right] + rv \right\} + \alpha\mathcal{L}_0 v = 0 \quad (10.4)$$

with $v(T, y) = 1$. The supremum is attained at

$$u^\star(t, y) = \frac{\beta\rho f(y)\frac{\partial v}{\partial y} + (\mu - r)v}{(1-p)f(y)^2 v}, \quad (10.5)$$

and we can rewrite the equation for v as

$$\frac{\partial v}{\partial t} + p\frac{\left(\beta\rho f(y)\frac{\partial v}{\partial y} + (\mu - r)v\right)^2}{2(1-p)f(y)^2 v} + \alpha\mathcal{L}_0 v + rpv = 0.$$

As expected, the optimal strategy requires knowledge of y.

As with our previous pricing and hedging problems, we take advantage of fast mean reversion in volatility and write

$$\alpha = 1/\varepsilon \quad \text{and} \quad \beta = \sqrt{2}v/\sqrt{\varepsilon},$$

and we suppose that ε is small. The equation then becomes

$$\frac{\partial v}{\partial t} + \frac{1}{\varepsilon}\left[\mathcal{L}_0 v + \frac{pv^2\rho^2}{(1-p)v}\left(\frac{\partial v}{\partial y}\right)^2 \right] + \frac{\sqrt{2}\rho vp(\mu - r)}{\sqrt{\varepsilon}(1-p)f(y)}\left(\frac{\partial v}{\partial y}\right)$$

$$+ \frac{p(\mu - r)^2 + 2p(1-p)rf(y)^2}{2(1-p)f(y)^2}v = 0. \quad (10.6)$$

Expanding v as

$$v = v_0 + \sqrt{\varepsilon}v_1 + \varepsilon v_2 + \varepsilon^{3/2}v_3 + \cdots$$

and comparing powers of ε, the term in $1/\varepsilon$ gives

$$\mathcal{L}_0 v_0 + \frac{pv^2\rho^2}{(1-p)v_0}\left(\frac{\partial v_0}{\partial y}\right)^2 = 0.$$

This is a differential equation in the y variable that reduces to

$$v^2\frac{v_0''}{v_0'} + (m - y) + \frac{pv^2\rho^2}{(1-p)}\frac{v_0'}{v_0} = 0,$$

where partial derivatives with respect to y are denoted by v_0' and v_0''. Integrating this expression yields

$$v_0^{1+c} = c_1(t)\int_0^y e^{(m-z)^2/2v^2}\, dz + c_2(t),$$

where $c = \rho^2 p/(1-p)$ is positive for our assumed $0 < p < 1$ and where c_1 and c_2 are some functions of time. These solutions do not belong to any reasonable space where the HJB equation is well-posed, unless $c_1(t) = 0$ and consequently v_0 does not depend on y. In other words, we must have $v_0 = v_0(t)$.

Comparing the $(1/\sqrt{\varepsilon})$-terms in the equation gives

$$\mathcal{L}_0 v_1 = 0,$$

because v_0 does not depend on y. This implies that $v_1 = v_1(t)$ is independent of y as well. This is an important observation, because it implies that the leading-order term in (10.5),

$$u^*(t, y)$$
$$= \frac{(\sqrt{2}v/\sqrt{\varepsilon})\rho f(y)\frac{\partial}{\partial y}(v_0 + \sqrt{\varepsilon}v_1 + \cdots) + (\mu - r)(v_0 + \sqrt{\varepsilon}v_1 + \cdots)}{(1-p)f(y)^2(v_0 + \sqrt{\varepsilon}v_1 + \cdots)},$$

is simply given by

$$u_0^*(y) = \frac{(\mu - r)}{(1-p)f(y)^2}, \tag{10.7}$$

which is independent of time and is the Merton solution (10.2) with σ^2 replaced by $f(y)^2$.

We can compute the leading term $v_0(t)$ of the maximum expected utility, which does not depend on y even though the optimal strategy does. Looking at the order-1 terms in (10.6), we have

$$\frac{\partial v_0}{\partial t} + \mathcal{L}_0 v_2 + \left(\frac{(\mu - r)^2}{2(1-p)f(y)^2} + r\right)pv_0 = 0, \tag{10.8}$$

which is a Poisson equation for v_2 with respect to $\mathcal{L}_0$. Its centering condition gives

$$\frac{\partial v_0}{\partial t} + \left(\frac{(\mu - r)^2}{2(1 - p)}\left\langle\frac{1}{f^2}\right\rangle + r\right)pv_0 = 0,$$

where $\langle\cdot\rangle$ again denotes the integral with respect to the invariant density of the OU process Y. If we set

$$\sigma^* = \frac{1}{\sqrt{\langle 1/f^2\rangle}} \tag{10.9}$$

then the leading-order maximum expected utility is exactly the Merton formula (10.3) with σ replaced by σ^*:

$$V(t, w, y) = \frac{w^p}{p}\exp\left(\left(r + \frac{(\mu - r)^2}{2(\sigma^*)^2(1 - p)}\right)p(T - t)\right) + \mathcal{O}(1/\sqrt{\alpha}).$$

We have seen that v_1 is also independent of y; it can be computed from the order-$\sqrt{\varepsilon}$ terms in (10.6), which give

$$\mathcal{L}_0 v_3 + \frac{\partial v_1}{\partial t} + \left(\frac{(\mu - r)^2}{2(1 - p)f(y)^2} + r\right)pv_1 + \frac{2pv\rho(\mu - r)}{(1 - p)f(y)}\frac{\partial v_2}{\partial y} = 0, \tag{10.10}$$

which is a Poisson equation in v_3. From (10.8) we deduce that

$$v_2 = -\frac{(\mu - r)^2 p}{2(1 - p)}(\phi^*(y) + k(t))v_0,$$

where $k(t)$ is some constant in y and ϕ^* satisfies

$$\mathcal{L}_0\phi^* = \frac{1}{f^2(y)} - \frac{1}{(\sigma^*)^2}.$$

The centering condition for (10.10) gives the following ordinary differential equation for $\tilde{v}_1 = (1/\sqrt{\alpha})v_1$:

$$\frac{\partial \tilde{v}_1}{\partial t} + \left(\frac{(\mu - r)^2}{2(1 - p)(\sigma^*)^2} + r\right)p\tilde{v}_1 = \frac{p^2(\mu - r)^3}{(1 - p)^2}A^*v_0,$$

where we define the small constant of order $\sqrt{\varepsilon}$ as

$$A^* = \frac{\rho v}{\sqrt{\alpha}}\left\langle\frac{(\phi^*)'}{f}\right\rangle;$$

this describes the correcting effect due to stochastic volatility. With the terminal condition $\tilde{v}_1(T) = 1$, we now calculate the corrected maximum expected utility as

$$V(t, w, y) = \frac{w^p}{p}\left(1 - \frac{p^2(\mu - r)^3 A^*}{(1 - p)^2}(T - t)\right)$$

$$\times \exp\left(\left(r + \frac{(\mu - r)^2}{2(\sigma^*)^2(1 - p)}\right)p(T - t)\right) + \mathcal{O}(1/\alpha).$$

Observe that, in this ideal scenario, the analysis identifies the market parameters σ^* and A^* as important statistics of the volatility. We also note that the transformation

$$v(t, y) = q^\delta(t, y)$$

with

$$\delta = \frac{1-p}{1-p+\rho^2 p}$$

leads to a *linear* parabolic partial differential equation for $q(t, y)$:

$$\frac{\partial q}{\partial t} + \frac{p(\mu - r)\beta\rho}{(1-p)f(y)}\left(\frac{\partial q}{\partial y}\right)$$
$$+ \frac{p(1-p+\rho^2 p)}{1-p}\left[r + \frac{(\mu - r)^2}{2f(y)^2(1-p)}\right]q + \alpha\mathcal{L}_0 q = 0.$$

Here δ is known as the distortion power. Alternatively, the asymptotic analysis can be performed on this equation (in a manner analogous to that described in Chapter 5). We give a reference for this transformation in the notes at the end of this chapter.

In practice, the volatility level (or, equivalently, Y_t) is not observable. The natural thing to do would be to filter Y_t from the observable price process X_t. This would lead to non-Markovian controls u depending on the past of X_t. This *filtering* problem is extremely complex, since Y_t appears in the "noise term" of the observation X_t. We will not pursue that route here, though performing the asymptotics on the filter is certainly worth investigating. Instead, we look at an intermediate solution that consists of restricting our set of strategies to those for which the leading-order term does not depend on volatility.

10.1.3   *A Practical Solution*

We now go back to the HJB equation (10.4) and restrict ourselves to strategies u that do not depend on the unobserved y: $u = u(t)$. As a result, the supremum in equation (10.4) is not well-defined unless the quantity to be maximized does not depend on y.

Expanding v as

$$v = v_0 + \sqrt{\varepsilon}v_1 + \cdots$$

and the control u as

$$u = u_0 + \sqrt{\varepsilon}u_1 + \cdots$$

and then collecting terms in (10.4), the $(1/\varepsilon)$-term implies

$$\mathcal{L}_0 v_0 = 0$$

and so $v_0 = v_0(t)$. Now, looking at the $(1/\sqrt{\varepsilon})$-terms, we have

$$\mathcal{L}_0 v_1 = 0$$

because v_0 does not depend on y. We deduce that $v_1 = v_1(t)$ also.

The order-1 terms in the equation give

$$\sup_{u_0}\left\{\frac{\partial v_0}{\partial t} + \left(\frac{1}{2}p(p-1)f(y)^2 u_0^2 + p(r + (\mu - r)u_0)\right)v_0 + \mathcal{L}_0 v_2\right\} = 0.$$

Since we are maximizing over u_0 independent of y, we must choose v_2 so that the argument of the supremum does not depend on y. If we define

$$A_{(f(y),u_0)} = \tfrac{1}{2}p(p-1)f(y)^2 u_0^2 + p(r + (\mu - r)u_0),$$

then the only way that the argument

$$\frac{\partial v_0}{\partial t} + A_{(f(y),u_0)}v_0 + \mathcal{L}_0 v_2$$

can be independent of y is by choosing v_2 to be a solution of the Poisson equation

$$\mathcal{L}_0 v_2 + (A_{(f(y),u_0)} - \langle A_{(f(y),u_0)}\rangle)v_0 = 0.$$

Now observe that, in terms of the function $\phi(y)$, the solution of the Poisson equation

$$\mathcal{L}_0 \phi = f^2(y) - \langle f^2\rangle$$

introduced in Section 5.2.3, v_2 is given by

$$v_2 = \tfrac{1}{2}p(1-p)u_0^2 v_0 \phi.$$

Consequently, the order-1 equation becomes

$$\frac{\partial v_0}{\partial t} + \sup_{u_0}\{\langle A_{(f(y),u_0)}\rangle)v_0\} = 0$$

with $v_0(T) = 1$, where

$$\langle A_{(f(y),u_0)}\rangle = A_{(\bar{\sigma},u_0)} = \tfrac{1}{2}p(p-1)\bar{\sigma}^2 u_0^2 + p(r + (\mu - r)u_0)$$

and $\bar{\sigma}^2 = \langle f^2\rangle$, the historical volatility.

Note that this is exactly the equation (10.1) with volatility $\bar{\sigma}$, which leads to the Merton solution

$$u_0^\star = \frac{(\mu - r)}{\bar{\sigma}^2(1 - p)}$$

and the corresponding maximum expected utility

$$V(t, w, y) = \frac{w^p}{p}\exp\left(\left(r + \frac{(\mu - r)^2}{2\bar{\sigma}^2(1 - p)}\right)p(T - t)\right) + \mathcal{O}(1/\sqrt{\alpha}).$$

From the inequality

$$\langle f^2 \rangle \left\langle \frac{1}{f^2} \right\rangle \geq \left\langle f \times \frac{1}{f} \right\rangle^2 = 1,$$

it follows that $\bar{\sigma} \geq \sigma^*$, where σ^* was defined in (10.9). In other words, as expected, the leading-order maximum expected utility would be greater if we could observe the volatility.

We now compute the correction v_1 by collecting the $\sqrt{\varepsilon}$-terms in (10.4). By the same argument, we can choose v_3 to make the quantity that is to be maximized at this order independent of y. Following steps analogous to the computation of v_1 in the previous section leads to an equation for the correction $\widetilde{v}_1 = (1/\sqrt{\alpha})v_1$:

$$\frac{\partial \widetilde{v}_1}{\partial t} + \left(\frac{(\mu - r)^2}{2\bar{\sigma}^2(1 - p)} + r \right) p \widetilde{v}_1 - \left(\frac{p}{1 - p} \right)^2 \left(\frac{(\mu - r)^3}{\bar{\sigma}^2} \right)^3 V_3 v_0 = 0,$$

where we have used

$$A_{(\bar{\sigma}, u_0^*)} = \left(\frac{(\mu - r)^2}{2\bar{\sigma}^2(1 - p)} + r \right) p$$

and where the small constant V_3 is

$$V_3 = \frac{\rho \nu}{\sqrt{2\alpha}} \langle f \phi' \rangle,$$

which is exactly the quantity in (5.40) whose estimation from the observed smile is discussed in Section 5.3. It is given by $V_3 = -a\bar{\sigma}^3$, where a is the slope of the fit to the smile (5.55).

We note also that the first correction to the strategy u_1 does not affect the first correction to the maximum expected utility. Using the formula for v_0, we obtain the corrected maximum expected utility:

$$V(t, w, y) = \frac{w^p}{p} \left[1 - \left(\frac{p}{1 - p} \right)^2 \left(\frac{\mu - r}{\bar{\sigma}^2} \right)^3 V_3(T - t) \right]$$

$$\times \exp\left(\left(r + \frac{(\mu - r)^2}{2\bar{\sigma}^2(1 - p)} \right) p(T - t) \right) + \mathcal{O}(1/\alpha).$$

The sign of the correction coefficient V_3 is opposite to that of the correlation ρ, which is typically negative. This shows that negative correlation diminishes the investor's maximum expected utility.

10.2   Periodic Day Effect

We have seen from the data analysis in Chapter 4 that volatility contains a daily periodic component. This is on the same scale as the intrinsic mean-reversion time of stochastic volatility. Therefore, it has to be included in the model, as in the simulations in Section 4.2.3. As we shall see here, the asymptotic results for pricing and hedging are not affected.

In order to model the daily effect, we replace our volatility model by $\sigma_t = f(\alpha t, Y_t)$, where $f(\tau, y)$ is positive and periodic of period 1 in its first argument. Recall that we found from data that $1/\alpha$ is on the order of a day and so, as a function of t, the volatility model contains the daily effect.

We return to the European pricing problem of Chapter 5. With the usual notation $\varepsilon = 1/\alpha$, it is convenient to introduce a new time variable $\tau = t/\varepsilon$ and consider the pricing function P^ε as a function $P^\varepsilon(t, \tau, x, y)$, treating τ as an independent variable. The usual time derivative is replaced by

$$\frac{\partial}{\partial t} + \frac{1}{\varepsilon}\frac{\partial}{\partial \tau}.$$

The pricing partial differential equation is now written

$$\mathcal{L}^\varepsilon P^\varepsilon = 0,$$

where, analogous to the definition in Section 5.1.3,

$$\mathcal{L}^\varepsilon = \frac{1}{\varepsilon}\left(\mathcal{L}_0 + \frac{\partial}{\partial \tau}\right) + \frac{1}{\sqrt{\varepsilon}}\mathcal{L}_1 + \mathcal{L}_0,$$

$$\mathcal{L}_0 = v^2\frac{\partial^2}{\partial y^2} + (m - y)\frac{\partial}{\partial y},$$

$$\mathcal{L}_1 = \sqrt{2}\rho v x f(\tau, y)\frac{\partial^2}{\partial x \partial y} - \sqrt{2}v\Lambda(\tau, y)\frac{\partial}{\partial y},$$

$$\mathcal{L}_2 = \frac{\partial}{\partial t} + \frac{1}{2}f(\tau, y)^2 x^2\frac{\partial^2}{\partial x^2} + r\left(x\frac{\partial}{\partial x} - \cdot\right) = \mathcal{L}_{\mathrm{BS}}(f(\tau, y)).$$

Notice that Λ depends on τ and that $\mathcal{L}_0$ is the usual scaled infinitesimal generator of (Y_t). The terminal condition is again $P^\varepsilon(T, \tau, x, y) = h(x)$.

Expanding as usual

$$P^\varepsilon = P_0 + \sqrt{\varepsilon}P_1 + \cdots$$

and comparing powers of ε gives, to highest order,

$$\left(\mathcal{L}_0 + \frac{\partial}{\partial \tau}\right)P_0 = 0.$$

The operator $\mathcal{L}_0 + \frac{\partial}{\partial \tau}$ is the infinitesimal generator of the process (Y_t, τ_t), where the second component is simply t modulo 1 to account for the periodic component.

Since $\mathcal{L}_0$ is independent of τ, the null space of $\mathcal{L}_0 + \frac{\partial}{\partial\tau}$ is made of the constants in (y, τ) and we deduce

$$P_0 = P_0(t, x).$$

The next order gives

$$\left(\mathcal{L}_0 + \frac{\partial}{\partial\tau}\right)P_1 = 0,$$

which again leads to a solution $P_1 = P_1(t, x)$. The zero-order terms yield the equation

$$\mathcal{L}_2 P_0 + \left(\mathcal{L}_0 + \frac{\partial}{\partial\tau}\right)P_2 = 0.$$

This equation has no solution unless $\mathcal{L}_2 P_0$ is centered with respect to the invariant measure of the Y process *and* over one period with respect to τ. We denote here by $\langle \cdot \rangle$ the integral

$$\langle g \rangle = \int_0^1 \int_{-\infty}^{\infty} g(\tau, y)\Phi(y)\, dy\, d\tau,$$

where Φ is the density of the invariant distribution of Y.

Using the fact that P_0 does not depend on τ or y, we have

$$\langle \mathcal{L}_2 \rangle P_0 = \mathcal{L}_{BS}(\bar\sigma) P_0 = 0,$$

where $\bar\sigma^2 = \langle f^2 \rangle$ and the average is taken in y and τ. In summary, the theory is the same with the modified averaging.

Following the argument given in Section 5.2.4, we obtain that the correction $\widetilde{P}_1(t, x) = \sqrt{\varepsilon}P_1(t, x)$ satisfies the equation

$$\mathcal{L}_{BS}(\bar\sigma)\widetilde{P}_1 = \sqrt{\varepsilon}\left\langle \mathcal{L}_1\left(\mathcal{L}_0 + \frac{\partial}{\partial\tau}\right)^{-1}(\mathcal{L}_2 - \langle\mathcal{L}_2\rangle)\right\rangle P_0,$$

with zero terminal condition. Introducing the function $\phi(\tau, y)$, a solution of

$$\left(\mathcal{L}_0 + \frac{\partial}{\partial\tau}\right)\phi(\tau, y) = f^2(\tau, y) - \langle f^2 \rangle,$$

the right side is computed as

$$V_2 x^2 \frac{\partial^2 P_0}{\partial x^2} + V_3 x^3 \frac{\partial^3 P_0}{\partial x^3},$$

where V_2 and V_3 are small constants given by the formulas (5.39) and (5.40) with the newly defined ϕ and $\langle \cdot \rangle$.

In conclusion, the use and calibration of the asymptotic theory is identical when we incorporate the day effect. Only the relations to the base model parameters are different.

10.3   Other Markovian Volatility Models

The purpose of this section is to explain how the analysis of Chapter 5 can be performed with volatility models other than the OU. That is, we illustrate the universality of the corrected pricing formula (5.43) among Markovian models. By a *general Markovian volatility model* we mean that the stock price process (X_t) evolves according to

$$dX_t = \mu X_t \, dt + f(Y_t) X_t \, dW_t,$$

where W is a standard Brownian motion, the volatility driving process (Y_t) is a Markov process, and the pair (X, Y) is Markovian. In the OU example considered in previous chapters, Y is defined by

$$dY_t = \alpha(m - Y_t) \, dt + \beta\big(\rho \, dW_t + \sqrt{1 - \rho^2} \, dZ_t\big),$$

where Z is a standard Brownian motion independent of W. The function f is a positive function bounded and bounded away from zero, and the leverage effect is built into the correlation ρ between the two Brownian motions driving the stock price and the volatility. A possible abstract setting could be a Markov process Y with jumps such that the leverage effect is created through the dependence between the Brownian motions driving the stock price and the diffusion component of Y. In this abstract context, the essential property of ergodicity is not easily related to the characteristics of the process Y.

   In order to avoid abstract settings, we keep the OU process and incorporate possible jumps in the volatility as another independent Markovian factor.

10.3.1   Markovian Jump Volatility Models

A convenient way to introduce jumps in volatility is to consider one of the examples of jump processes given in Sections 3.2.1 or 3.2.2, call it (ξ_t), and model the stochastic volatility σ_t as a function

$$\sigma_t = f(Y_t, \xi_t)$$

of the OU process Y and the process ξ, which we assume to be independent. For instance, if f is a function only of ξ, then the volatility is a pure jump process independent of W and there is no leverage effect. We are mostly interested in cases where the two components Y and ξ are present. This procedure corresponds to the addition of another *factor,* which models the volatility jumps. We also assume that the intensity of the jumps is large and of the same order as the rate of mean reversion of the OU process Y. This is done by setting this intensity equal to the rate of mean reversion of Y, namely α.

The properties of the Markov process (Y, ξ) just described can be summarized in its infinitesimal generator $\mathcal{L}_{(Y,\xi)}$, given by

$$\mathcal{L}_{(Y,\xi)} = \alpha(\mathcal{L}_0 + \mathcal{L}_J),$$

where $\mathcal{L}_0$ is the infinitesimal generator of the OU process acting on the variable y,

$$\mathcal{L}_0 = \nu^2 \frac{\partial^2}{\partial y^2} + (m - y)\frac{\partial}{\partial y},$$

and where $\mathcal{L}_J$, acting on the second variable (also denoted by ξ), is the infinitesimal generator of one of the jump processes described in Sections 3.2.1 or 3.2.2, with an intensity normalized to 1. For instance, in the case of jumps to a random point in $[-1, 1]$ (as considered in Section 3.2.2), we have

$$\frac{1}{\alpha}\mathcal{L}_{(Y,\xi)}g(y, \xi) = \nu^2 \frac{\partial^2 g}{\partial y^2}(y, \xi) + (m - y)\frac{\partial g}{\partial y}(y, \xi)$$

$$+ \frac{1}{2}\int_{-1}^{1} (g(y, z) - g(y, \xi))\, dz \qquad (10.11)$$

for any bounded function g that is twice differentiable with respect to y.

An important feature of this setting is that the process (Y, ξ) is ergodic and admits the unique invariant distribution described by

$$\langle g \rangle = \langle\langle g \rangle_Y \rangle_\xi = \langle\langle g \rangle_\xi \rangle_Y, \qquad (10.12)$$

which consists of averaging in the first variable with respect to the invariant distribution of the OU process Y and in the second variable with respect to the invariant distribution of the process ξ. For instance, in the example (10.11) we have

$$\langle g \rangle = \frac{1}{2\nu\sqrt{2\pi}}\int_{-1}^{1}\int_{-\infty}^{+\infty} g(y, z)\exp\left(-\frac{(y - m)^2}{2\nu^2}\right) dy\, dz.$$

Using the single small parameter $\varepsilon = 1/\alpha$, the generalized Markovian model $(X^\varepsilon, Y^\varepsilon, \xi^\varepsilon)$ is given, in the subjective world $\mathbb{P}$, by

$$dX_t^\varepsilon = \mu X_t^\varepsilon\, dt + f(Y_t^\varepsilon, \xi_t^\varepsilon)X_t^\varepsilon\, dW_t,$$

$$dY_t^\varepsilon = \frac{1}{\varepsilon}(m - Y_t^\varepsilon)\, dt + \frac{\nu\sqrt{2}}{\sqrt{\varepsilon}}(\rho\, dW_t + \sqrt{1 - \rho^2}\, dZ_t), \qquad (10.13)$$

$$\xi_t^\varepsilon = \xi_{t/\varepsilon},$$

where (W, Z, ξ) are independent. We assume that the factor ξ remains the same in the risk-neutral world $\mathbb{P}^\star$, so that

$$dX_t^\varepsilon = rX_t^\varepsilon \, dt + f(Y_t^\varepsilon, \xi_t^\varepsilon) X_t^\varepsilon \, dW_t^\star,$$

$$dY_t^\varepsilon = \frac{1}{\varepsilon}(m - Y_t^\varepsilon) \, dt - \frac{v\sqrt{2}}{\sqrt{\varepsilon}} \Lambda(Y_t^\varepsilon, \xi_t^\varepsilon) \, dt$$

$$+ \frac{v\sqrt{2}}{\sqrt{\varepsilon}} \left(\rho \, dW_t^\star + \sqrt{1 - \rho^2} \, dZ_t^\star \right),$$

$$\xi_t^\varepsilon = \xi_{t/\varepsilon},$$

$$(10.14)$$

where $(W^\star, Z^\star, \xi)$ are independent and Λ may also depend on ξ_t^ε.

10.3.2   *Pricing and Asymptotics*

In this setting, the price of a European derivative with payoff h at maturity T is given at time $t < T$ by $P^\varepsilon(t, X_t^\varepsilon, Y_t^\varepsilon, \xi_t^\varepsilon)$, where

$$P^\varepsilon(t, x, y, \xi) = I\!E^\star\{e^{-r(T-t)}h(X_T^\varepsilon) \mid X_t^\varepsilon = x, \ Y_t^\varepsilon = y, \ \xi_t^\varepsilon = \xi\}$$

satisfies

$$\left(\frac{1}{\varepsilon}(\mathcal{L}_0 + \mathcal{L}_J) + \frac{1}{\sqrt{\varepsilon}}\mathcal{L}_1 + \mathcal{L}_{BS}(f(y, \xi)) \right) P^\varepsilon = 0$$

with the terminal condition $P^\varepsilon(T, x, y, \xi) = h(x)$. The operator $\mathcal{L}_0$ is the usual OU infinitesimal generator acting on y; $\mathcal{L}_J$ is the infinitesimal generator of the jump process acting on ξ; $\mathcal{L}_1$ is given as in (5.12) by

$$\mathcal{L}_1 = v\sqrt{2}\rho x f(y, \xi) \frac{\partial^2}{\partial x \partial y} - v\sqrt{2}\Lambda(y, \xi) \frac{\partial}{\partial y}$$

with Λ, which may also depend on ξ; and $\mathcal{L}_{BS}(f(y, \xi))$ is the Black–Scholes operator with volatility $f(y, \xi)$.

With these definitions and notation, the rest of the derivation of the **corrected pricing formula** (5.43),

$$P_0 - (T - t)\left(V_2 x^2 \frac{\partial^2 P_0}{\partial x^2} + V_3 x^3 \frac{\partial^3 P_0}{\partial x^3} \right),$$

follows the lines of Section 5.2. The function $P_0(t, x)$ is the Black–Scholes price of the derivative, computed with the constant volatility $\bar{\sigma}$ defined according to (10.12) by

$$\bar{\sigma}^2 = \langle f^2 \rangle_{(Y, \xi)}.$$

The parameters V_2 and V_3 are obtained by generalizing the formulas (5.39) and (5.40):

$$V_2 = \frac{\nu}{\sqrt{2\alpha}} \left(2\rho \left\langle f \frac{\partial \phi}{\partial y} \right\rangle - \left\langle \Lambda \frac{\partial \phi}{\partial y} \right\rangle \right),$$

$$V_3 = \frac{\rho\nu}{\sqrt{2\alpha}} \left\langle f \frac{\partial \phi}{\partial y} \right\rangle,$$

where the averages are taken with respect to the invariant distribution of (Y, ξ), and $\phi(y, \xi)$ is a solution of the **Poisson equation**

$$(\mathcal{L}_0 + \mathcal{L}_J)\phi(y, \xi) = f(y, \xi)^2 - \bar{\sigma}^2. \tag{10.15}$$

The main assumption on the added factor ξ is that this equation admits well-behaved solutions as found for (5.22). This is the case for bounded processes (cf. the examples considered in Sections 3.2.1 or 3.2.2), where a solution can be written as in (5.25),

$$\phi(y, \xi) = \int_0^{+\infty} I\!\!E\{(f(Y_t, \xi_t)^2 - \bar{\sigma}^2) \mid Y_0 = y, \ \xi_0 = \xi\} \, dt.$$

The validity – in this generalized Markovian context – of the corrected pricing formula (5.43) illustrates its **universality**. It can be computed from the Black–Scholes price $P_0(t, x)$, the Gamma $\partial^2 P_0/\partial x^2$, and the Epsilon $\partial^3 P_0/\partial x^3$, and it can be calibrated on the implied volatility surface by using (5.55)–(5.57) without any further knowledge of the processes (Y_t) and (ξ_t) or of their current values y and ξ.

10.4 Martingale Approach

The goal of this section is to show that the approximation of the price of a European derivative obtained in Chapter 5 can also be derived by using a martingale approach. This will be useful in handling non-Markovian models, as we shall explain in Section 10.5. The martingale approach can also be used for American derivatives by using the supermartingale characterization of their prices, but this requires more details on the optimal stopping theory and will be treated elsewhere.

10.4.1 Main Argument

In order to explain the martingale approach, we develop it first in the now-familiar OU Markovian context where, in the risk-neutral world $I\!\!P^\star$, the pair $(X_t^\varepsilon, Y_t^\varepsilon)$ satisfies the system of stochastic differential equations (5.4)–(5.5). Their infinitesimal generator is given by

$$L^\varepsilon = \frac{1}{\varepsilon}\mathcal{L}_0 + \frac{1}{\sqrt{\varepsilon}}\mathcal{L}_1 + \frac{1}{2}f(y)^2 x^2 \frac{\partial^2}{\partial x^2} + rx\frac{\partial}{\partial x}$$

(with the notation of Section 5.1.3); $\mathcal{L}_0$ is the infinitesimal generator of the driving OU process Y, normalized to a unit rate of mean reversion:

$$\mathcal{L}_0 = v^2 \frac{\partial^2}{\partial y^2} + (m - y) \frac{\partial}{\partial y}.$$

Also,

$$\mathcal{L}_1 = \sqrt{2} \rho v x f(y) \frac{\partial^2}{\partial x \partial y} - \sqrt{2} v \Lambda(y) \frac{\partial}{\partial y}.$$

From Section 2.5 we know that the price $P^\varepsilon(t)$ at time t of a European derivative with terminal payoff function h is given by

$$P^\varepsilon(t) = I\!E^\star \{ e^{-r(T-t)} h(X_T^\varepsilon) \mid \mathcal{F}_t \},$$

where the conditional expectation is taken under the equivalent martingale measure $I\!P^\star$ chosen by the market and with respect to the past of $(X^\varepsilon, Y^\varepsilon)$ up to time t, denoted by $\mathcal{F}_t$. Note that this is also the natural filtration of the Brownian motions (W, Z). The price $P^\varepsilon(t)$ is also characterized by the fact that the process M^ε defined by

$$M_t^\varepsilon = e^{-rt} P^\varepsilon(t) = I\!E^\star \{ e^{-rT} h(X_T^\varepsilon) \mid \mathcal{F}_t \} \tag{10.16}$$

is a martingale, with a terminal value given by

$$M_T^\varepsilon = e^{-rT} h(X_T^\varepsilon).$$

On the other hand, for a given function $Q^\varepsilon(t, x)$ that may depend on ε but not on y, we consider the process N^ε defined by

$$N_t^\varepsilon = e^{-rt} Q^\varepsilon(t, X_t^\varepsilon). \tag{10.17}$$

If the function Q^ε satisfies $Q^\varepsilon(T, x) = h(x)$ at the final time T, then

$$M_T^\varepsilon = N_T^\varepsilon. \tag{10.18}$$

The method consists of finding $Q^\varepsilon(t, x)$ such that N^ε can be **decomposed** as

$$N_t^\varepsilon = \widetilde{M_t^\varepsilon} + R_t^\varepsilon, \tag{10.19}$$

where $\widetilde{M^\varepsilon}$ is a martingale and R_t^ε is of order ε. Observe that, in this decomposition, we require that the terms of order $\sqrt{\varepsilon}$ be absorbed into the martingale part.

Suppose that this has been established. By taking a conditional expectation with respect to $\mathcal{F}_t$ on both sides of the equality

$$N_T^\varepsilon = \widetilde{M_T^\varepsilon} + R_T^\varepsilon,$$

we have

$$I\!E^\star \{ N_T^\varepsilon \mid \mathcal{F}_t \} = \widetilde{M_t^\varepsilon} + I\!E^\star \{ R_T^\varepsilon \mid \mathcal{F}_t \},$$

by the martingale property of $\widetilde{M}^\varepsilon$. From (10.18) and the martingale property of M^ε we deduce that the left-hand side is also equal to M_t^ε. From the decomposition (10.19) we have $\widetilde{M}_t^\varepsilon = N_t^\varepsilon - R_t^\varepsilon$ and so

$$M_t^\varepsilon = N_t^\varepsilon + I\!E^\star\{R_T^\varepsilon \mid \mathcal{F}_t\} - R_t^\varepsilon.$$

Multiplying by e^{rt} and using (10.16) and (10.17) yields

$$P^\varepsilon(t) = Q^\varepsilon(t, X_t^\varepsilon) + \mathcal{O}(\varepsilon),$$

which is the desired **approximation result**.

It remains to determine Q^ε leading to the decomposition (10.19) of N^ε. The technique is basically the same as used in Chapter 7 and explained in Section 7.1.2; the only difference is that the computation is done in the risk-neutral world with the process $(X^\varepsilon, Y^\varepsilon)$. We outline the technique in the next section and show that Q^ε can be chosen equal to $P_0 + \sqrt{\varepsilon}P_1 = P_0 + \widetilde{P}_1$, where P_0 and $\widetilde{P}_1$ have been obtained in Chapter 5.

10.4.2 Decomposition Result

From the definition (10.17) of N^ε and the fact that Q^ε depends only on (t, x), we deduce that

$$dN_t^\varepsilon = e^{-rt}\left(\frac{\partial}{\partial t} + \frac{1}{2}f(Y_t^\varepsilon)^2(X_t^\varepsilon)^2\frac{\partial^2}{\partial x^2} + rX_t^\varepsilon\frac{\partial}{\partial x} - r\right)Q^\varepsilon(t, X_t^\varepsilon)\,dt$$

$$+ e^{-rt}\left(\frac{\partial Q^\varepsilon}{\partial x}(t, X_t^\varepsilon)\right)f(Y_t^\varepsilon)X_t^\varepsilon\,dW_t^\star, \tag{10.20}$$

where we assume a priori sufficient smoothness on the function Q^ε.

The Markov process Y^ε given by (5.5) admits the infinitesimal generator $\varepsilon^{-1}\mathcal{L}_0^\varepsilon$, where

$$\mathcal{L}_0^\varepsilon = \mathcal{L}_0 - v\sqrt{2\varepsilon}\Lambda(y)\frac{\partial}{\partial y}$$

$$= v^2\frac{\partial^2}{\partial y^2} + (m - y - v\sqrt{2\varepsilon}\Lambda(y))\frac{\partial}{\partial y} \tag{10.21}$$

is a perturbation of the OU infinitesimal generator $\mathcal{L}_0$. Assuming that $|\Lambda(y)|$ is bounded, Y^ε has a unique invariant distribution that is a perturbation of the invariant distribution of the OU process Y given by the probability density

$$\Phi_\varepsilon(y) = J_\varepsilon \exp\left(-\frac{(y - m)^2}{2v^2} - \frac{\sqrt{2\varepsilon}}{v}\widetilde{\Lambda}(y)\right), \tag{10.22}$$

where $\widetilde{\Lambda}$ is an antiderivative of Λ that is at most linear at infinity and J_ε is a normalization constant depending on ε. This is easily seen by first writing

$$\mathcal{L}_0^\varepsilon = v^2 \exp\left(\left\{\frac{(y-m)^2}{2v^2} + \frac{\sqrt{2\varepsilon}}{v}\widetilde{\Lambda}(y)\right\}\right)\frac{\partial}{\partial y}$$

$$\times \left(\exp\left(-\left\{\frac{(y-m)^2}{2v^2} + \frac{\sqrt{2\varepsilon}}{v}\widetilde{\Lambda}(y)\right\}\right)\frac{\partial \cdot}{\partial y}\right)$$

$$= \frac{v^2}{\Phi_\varepsilon}((\Phi_\varepsilon \cdot)')', \tag{10.23}$$

then deducing the adjoint operator of $\mathcal{L}_0^\varepsilon$ (using integration by parts),

$$(\mathcal{L}_0^\varepsilon)^\star = v^2\left(\Phi_\varepsilon\left(\frac{\cdot}{\Phi_\varepsilon}\right)'\right)',$$

and concluding that Φ_ε given by (10.22) satisfies

$$(\mathcal{L}_0^\varepsilon)^\star \Phi_\varepsilon = 0.$$

We will denote by $\langle \cdot \rangle_\varepsilon$ the expectation with respect to this invariant distribution Φ_ε,

$$\langle g \rangle_\varepsilon = J_\varepsilon \int g(y)\exp\left(-\frac{(y-m)^2}{2v^2} - \frac{\sqrt{2\varepsilon}}{v}\widetilde{\Lambda}(y)\right)dy,$$

and define the constant volatility $\bar{\sigma}_\varepsilon$ by

$$\bar{\sigma}_\varepsilon^2 = \langle f^2 \rangle_\varepsilon. \tag{10.24}$$

We rewrite equation (10.20) by introducing the Black–Scholes operator $\mathcal{L}_{\mathrm{BS}}(\bar{\sigma}_\varepsilon)$:

$$dN_t^\varepsilon = e^{-rt}\left(\mathcal{L}_{\mathrm{BS}}(\bar{\sigma}_\varepsilon) + \frac{1}{2}(f(Y_t^\varepsilon)^2 - \bar{\sigma}_\varepsilon^2)(X_t^\varepsilon)^2\frac{\partial^2}{\partial x^2}\right)Q^\varepsilon(t, X_t^\varepsilon)\,dt$$

$$+ e^{-rt}\left(\frac{\partial Q^\varepsilon}{\partial x}(t, X_t^\varepsilon)\right)f(Y_t^\varepsilon)X_t^\varepsilon\,dW_t^\star. \tag{10.25}$$

From this point, the technique is identical to what we have done in Chapter 7 except that $\mathcal{L}_{\mathrm{BS}}(\bar{\sigma})$ is replaced by $\mathcal{L}_{\mathrm{BS}}(\bar{\sigma}_\varepsilon)$ and the averages are with respect to Φ_ε instead of Φ. We make the following definitions.

(1) The solution $P_0^\varepsilon(t, x)$ of the Black–Scholes equation

$$\mathcal{L}_{\mathrm{BS}}(\bar{\sigma}_\varepsilon)P_0^\varepsilon = 0 \tag{10.26}$$

with the terminal condition $P_0^\varepsilon(T, x) = h(x)$.

(2) A solution ϕ_ε of the Poisson equation

$$\mathcal{L}_0^\varepsilon \phi_\varepsilon(y) = f(y)^2 - \bar{\sigma}_\varepsilon^2, \tag{10.27}$$

obtained as in (5.31). Using (10.23), one can deduce

$$\phi'_\varepsilon = \frac{1}{v^2 \Phi_\varepsilon} \int_{-\infty}^{\cdot} (f^2 - \langle f^2 \rangle_\varepsilon) \Phi_\varepsilon,$$

as well as estimates similar to (5.26).

(3) The quantity

$$V_3^\varepsilon = \frac{\rho v \sqrt{\varepsilon}}{\sqrt{2}} \langle f \phi'_\varepsilon \rangle_\varepsilon. \tag{10.28}$$

(4) The source term

$$H^\varepsilon(t, x) = V_3^\varepsilon x \frac{\partial}{\partial x} \left(x^2 \frac{\partial^2 P_0^\varepsilon}{\partial x^2} \right) = V_3^\varepsilon \left(2x^2 \frac{\partial^2 P_0^\varepsilon}{\partial x^2} + x^3 \frac{\partial^3 P_0^\varepsilon}{\partial x^3} \right), \tag{10.29}$$

similar to the source term used in Section 7.2.

(5) The solution $\widetilde{Q}_1^\varepsilon(t, x)$ of the Black–Scholes equation

$$\mathcal{L}_{BS}(\bar{\sigma}_\varepsilon) \widetilde{Q}_1^\varepsilon = H^\varepsilon, \tag{10.30}$$

with the zero terminal condition $\widetilde{Q}_1^\varepsilon(T, x) = 0$.

(6) The function $Q^\varepsilon(t, x)$ is defined by

$$Q^\varepsilon = P_0^\varepsilon + \widetilde{Q}_1^\varepsilon \tag{10.31}$$

and satisfies the terminal condition $Q^\varepsilon(T, x) = h(x)$ necessary to (10.18).

For this choice of Q^ε, equation (10.25) can be rewritten as

$$dN_t^\varepsilon = e^{-rt} \left(H^\varepsilon + \frac{1}{2} (f(Y_t^\varepsilon)^2 - \bar{\sigma}_\varepsilon^2)(X_t^\varepsilon)^2 \frac{\partial^2 Q^\varepsilon}{\partial x^2} \right) (t, X_t^\varepsilon) \, dt$$

$$+ e^{-rt} \left(\frac{\partial Q^\varepsilon}{\partial x} (t, X_t^\varepsilon) \right) f(Y_t^\varepsilon) X_t^\varepsilon \, dW_t^\star. \tag{10.32}$$

Following the technique described in Section 7.1.2, we deduce from Itô's formula and the definition (10.27) of the function ϕ_ε that

$$(f(Y_t^\varepsilon)^2 - \bar{\sigma}_\varepsilon^2) \, dt = \varepsilon \left\{ d(\phi_\varepsilon(Y_t^\varepsilon)) - \frac{v\sqrt{2}}{\sqrt{\varepsilon}} \phi'_\varepsilon(Y_t^\varepsilon) \, d\hat{Z}_t^\star \right\}. \tag{10.33}$$

Using the integration-by-parts formula and following the lines of Section 7.2, the H^ε-term is exactly designed to center the $(\sqrt{\varepsilon} \, dt)$-terms. This leads to

$$N_t^\varepsilon = N_0^\varepsilon + \int_0^t e^{-rs} \left(\frac{\partial Q^\varepsilon}{\partial x} (s, X_s^\varepsilon) \right) f(Y_s^\varepsilon) X_s^\varepsilon \, dW_s^\star$$

$$- \frac{v\sqrt{\varepsilon}}{\sqrt{2}} \int_0^t (X_s^\varepsilon)^2 \frac{\partial^2 P_0^\varepsilon}{\partial x^2} (s, X_s^\varepsilon) \phi'_\varepsilon(Y_s^\varepsilon) \, d\hat{Z}_s^\star + \mathcal{O}(\varepsilon),$$

which is the desired decomposition (10.19) of N^ε by defining the martingale

$$\widetilde{M}_t^\varepsilon = N_0^\varepsilon + \int_0^t e^{-rs} \left(\frac{\partial Q^\varepsilon}{\partial x}(s, X_s^\varepsilon) \right) f(Y_s^\varepsilon) X_s^\varepsilon \, dW_s^\star$$

$$- \frac{\nu\sqrt{\varepsilon}}{\sqrt{2}} \int_0^t (X_s^\varepsilon)^2 \frac{\partial^2 P_0^\varepsilon}{\partial x^2}(s, X_s^\varepsilon) \phi_\varepsilon'(Y_s^\varepsilon) \, d\hat{Z}_s^\star$$

(which absorbs the order-1 and order-$\sqrt{\varepsilon}$ terms) and collecting all the remaining terms, which are of order ε and constitute the process R^ε in (10.19).

10.4.3 Comparison with the PDE Approach

We show in this section how to derive, from $Q^\varepsilon(t, x)$ given by (10.31), the approximate price $P_0(t, x) + \widetilde{P}_1(t, x)$ that we obtained in Chapter 5 by working directly on the partial differential equation satisfied by the modeled price $P(t, x, y)$. These two approximations are functions of (t, x) only; they have the same order of approximation $\mathcal{O}(\varepsilon) = \mathcal{O}(1/\alpha)$; and they have the same terminal condition $h(x)$.

By expanding (10.22) with respect to $\sqrt{\varepsilon}$, including the normalization constant J_ε, it is easily seen that the invariant distribution of Y^ε is related to the invariant distribution of Y by

$$\langle \cdot \rangle_\varepsilon = \langle \cdot \rangle - \frac{\sqrt{2\varepsilon}}{\nu} \langle \widetilde{\Lambda}(\cdot - \langle \cdot \rangle) \rangle + \mathcal{O}(\varepsilon). \qquad (10.34)$$

Therefore,

$$\mathcal{L}_{BS}(\bar{\sigma}_\varepsilon) = \mathcal{L}_{BS}(\bar{\sigma}) + \frac{1}{2}(\langle f^2 \rangle_\varepsilon - \langle f^2 \rangle) x^2 \frac{\partial^2}{\partial x^2}$$

$$= \mathcal{L}_{BS}(\bar{\sigma}) - \frac{\sqrt{\varepsilon}}{\nu\sqrt{2}} \langle \widetilde{\Lambda}(f^2 - \langle f^2 \rangle) \rangle x^2 \frac{\partial^2}{\partial x^2} + \mathcal{O}(\varepsilon)$$

$$= \mathcal{L}_{BS}(\bar{\sigma}) + \frac{\nu\sqrt{\varepsilon}}{\sqrt{2}} \langle \Lambda \phi' \rangle x^2 \frac{\partial^2}{\partial x^2} + \mathcal{O}(\varepsilon), \qquad (10.35)$$

according to the calculation made in Section 5.2.5. We see here that the market price of risk Λ induces a correction to the mean historical volatility $\bar{\sigma}$. The source term induced by the leverage effect is of a different nature, as we observed in Section 5.2.8.

Replacing ϕ_ε by ϕ and $\langle \cdot \rangle_\varepsilon$ by $\langle \cdot \rangle$ in (10.28), we can replace V_3^ε by V_3 in (10.29) without affecting the $\sqrt{\varepsilon}$-order term. Still without affecting this order, we can then replace P_0^ε by P_0, the solution of the Black–Scholes equation with constant volatility $\bar{\sigma}$. Putting together (10.26), (10.35), and (10.30), we obtain that Q^ε defined by (10.31) solves

$$\mathcal{L}_{\mathrm{BS}}(\bar{\sigma})Q^{\varepsilon} = V_3\left(2x^2\frac{\partial^2 P_0}{\partial x^2} + x^3\frac{\partial^3 P_0}{\partial x^3}\right) - \frac{\nu\sqrt{\varepsilon}}{\sqrt{2}}\langle\Lambda\phi'\rangle x^2\frac{\partial^2 P_0}{\partial x^2} + \mathcal{O}(\varepsilon)$$

$$= V_2 x^2\frac{\partial^2 P_0}{\partial x^2} + V_3 x^3\frac{\partial^3 P_0}{\partial x^3} + \mathcal{O}(\varepsilon),$$

according to the definition (5.39) of the parameter V_2. Except for the terms of order ε collected in $\mathcal{O}(\varepsilon)$, this is the same equation defining $P_0 + \widetilde{P}_1$ that was obtained in Chapter 5, since the sources and the terminal conditions are the same. Hence, Q^{ε} and $P_0 + \widetilde{P}_1$ differ only at the order ε.

Observe that in practice we will use $P_0 + \widetilde{P}_1$ and so require only the parameters $(\bar{\sigma}, V_2, V_3)$, which are easily calibrated (as summarized in Chapter 6). Nevertheless, the martingale approach presented here shows the universality of the procedure and will be fully exploited in the non-Markovian situations presented in the next section.

10.5 Non-Markovian Models of Volatility

In this section we indicate how the martingale approach presented in Section 10.4 can be used to handle non-Markovian models of stochastic volatility. Writing the most general model that would fit within our theory would be much too technical. Instead, as in Section 10.3, we build a class of models on top of the OU model. We consider volatility driving processes $\widetilde{Y}_t = (Y_t, \xi_t)$, where the additional factor is an independent ergodic process (ξ_t) with a *rate of mixing* of the same order as the rate of mean reversion of Y. The complete model is given by (10.13) in the subjective world and by (10.14) in the risk-neutral world.

10.5.1 Setting: An Example

In order to illustrate this setting, we consider a concrete example where Y is the OU process and ξ is an independent stationary centered Gaussian process with covariance denoted by Γ:

$$\mathbb{E}\{\xi_s\xi_{s+t}\} = \Gamma(t).$$

Unless $\Gamma(t)$ is proportional to $e^{-a|t|}$ for some constant $a > 0$ (in which case ξ would be also an OU process), in general ξ is not Markovian. A convenient way to write such a process is by using its spectral representation

$$\xi_t = \int e^{i\omega t}\sqrt{2\pi\hat{\Gamma}(\omega)}\widetilde{W}(d\omega), \tag{10.36}$$

where $\widetilde{W}$ is a *Gaussian white noise* in the frequency domain, independent of the Brownian motions driving X and Y, and $\hat{\Gamma}$ is the Fourier transform of the covariance function. We have

$$\mathbb{E}\{\xi_s \xi_{s+t}\} = 2\pi \iint e^{i\omega s} e^{-i\omega'(s+t)} \sqrt{\hat{\Gamma}(\omega)\hat{\Gamma}(\omega')} \, \mathbb{E}\{\widetilde{W}(d\omega)\overline{\widetilde{W}(d\omega')}\}$$

$$= 2\pi \int e^{-i\omega t} \hat{\Gamma}(\omega) \, d\omega$$

$$= \Gamma(t),$$

where the decay at infinity of Γ is controlled by the behavior of its Fourier transform $\hat{\Gamma}$ at low frequencies, which we suppose are smooth.

An interesting particular case is obtained by considering functions $f(y, \xi)$ and $\Lambda(y, \xi)$ that depend on the sum $y + \xi$ only. In that case, the volatility σ_t is driven by the Gaussian process $\widetilde{Y}^\varepsilon = Y^\varepsilon + \xi^\varepsilon$ and, under the risk-neutral probability $\mathbb{P}^\star$ chosen by the market, the model is now as follows:

$$dX_t^\varepsilon = rX_t^\varepsilon \, dt + f(\widetilde{Y}_t^\varepsilon) X_t^\varepsilon \, dW_t^\star,$$

$$\widetilde{Y}_t^\varepsilon = Y_t^\varepsilon + \xi_t^\varepsilon,$$

$$dY_t^\varepsilon = \left[\frac{1}{\varepsilon}(m - Y_t^\varepsilon) - \frac{\nu\sqrt{2}}{\sqrt{\varepsilon}} \Lambda(\widetilde{Y}_t^\varepsilon) \right] dt \qquad (10.37)$$

$$+ \frac{\nu\sqrt{2}}{\sqrt{\varepsilon}} (\rho \, dW_t^\star + \sqrt{1-\rho^2} \, dZ_t^\star),$$

$$\xi_t^\varepsilon = \xi_{t/\varepsilon} = \int e^{i\omega t/\varepsilon} \sqrt{2\pi \hat{\Gamma}(\omega)} \, \widetilde{W}(d\omega),$$

where $W^\star$ and $Z^\star$ are two independent standard Brownian motions that are independent of the white noise $\widetilde{W}$.

We denote by $\langle g \rangle_\varepsilon$ the average of a function $g(y, \xi)$ with respect to the invariant distribution of the process $(Y^\varepsilon, \xi^\varepsilon)$,

$$\langle g \rangle_\varepsilon = \iint g(y, \xi) \Phi_\varepsilon(y, \xi) p_\Gamma(\xi) \, dy \, d\xi, \qquad (10.38)$$

where Φ_ε is the density of the perturbed OU given by (10.22), which may now depend on ξ through $\Lambda(y, \xi)$, and p_Γ is the $\mathcal{N}(0, \Gamma(0))$-density of the Gaussian invariant distribution of the process (ξ_t).

10.5.2 Asymptotics in the Non-Markovian Case

The major difference here from the Markovian models considered previously is that no-arbitrage prices of derivatives are not obtained as solutions of partial differential equations, since we have "lost" the Markov property and the associated tools (e.g., infinitesimal generators). These prices are obtained as conditional expectations with respect to the past of $(X^\varepsilon, Y^\varepsilon, \xi^\varepsilon)$ defined in (10.37). This filtration will be denoted by $(\mathcal{F}_t^\varepsilon)$. For example, the price at time t of a European derivative contract with payoff function h at maturity T is given by

$$P^\varepsilon(t) = I\!E^\star\{e^{-r(T-t)}h(X_T^\varepsilon) \mid \mathcal{F}_t^\varepsilon\}$$

or characterized by the fact that

$$M_t^\varepsilon = e^{-rt}P^\varepsilon(t)$$

is a martingale with respect to $(\mathcal{F}_t^\varepsilon)$ with terminal value $M_T^\varepsilon = h(X_T^\varepsilon)$.

As in Section 10.4.1, we seek a function $Q^\varepsilon(t, x)$ satisfying the terminal condition

$$Q^\varepsilon(T, x) = h(x)$$

and such that the process N^ε defined by

$$N_t^\varepsilon = e^{-rt}Q^\varepsilon(t, X_t^\varepsilon)$$

admits the decomposition (10.19),

$$N_t^\varepsilon = \widetilde{M_t^\varepsilon} + R_t^\varepsilon,$$

where $\widetilde{M^\varepsilon}$ is a martingale absorbing the order $\sqrt{\varepsilon}$ and R_t^ε is of order ε. Following the martingale argument given in Section 10.4.1, one can easily deduce the approximation result

$$P^\varepsilon(t) = Q^\varepsilon(t, X_t^\varepsilon) + \mathcal{O}(\varepsilon),$$

where it remains to construct such a function Q^ε.

In the setting of Section 10.5.1, we can still write equations similar to (10.20) and (10.25) for dN_t^ε by replacing Y^ε with $(Y^\varepsilon, \xi^\varepsilon)$. This gives

$$dN_t^\varepsilon = e^{-rt}\left(\mathcal{L}_{BS}(\bar{\sigma}_\varepsilon) + \frac{1}{2}(f(Y_t^\varepsilon, \xi_t^\varepsilon)^2 - \bar{\sigma}_\varepsilon^2)(X_t^\varepsilon)^2 \frac{\partial^2}{\partial x^2}\right)Q^\varepsilon(t, X_t^\varepsilon)\,dt$$

$$+ e^{-rt}\left(\frac{\partial Q^\varepsilon}{\partial x}(t, X_t^\varepsilon)\right)f(Y_t^\varepsilon, \xi_t^\varepsilon)X_t^\varepsilon\,dW_t^\star, \tag{10.39}$$

where $\bar{\sigma}_\varepsilon^2$ is the average (10.38) of f^2.

We then define, as in (10.26), the solution $P_0^\varepsilon(t, x)$ of the Black–Scholes equation with constant volatility $\bar{\sigma}_\varepsilon$. Because we are not in a Markovian setting, we can no longer write the Poisson equation defining the function $\phi_\varepsilon(y, \xi)$ as in (10.15) and (10.27). Instead, $\phi_\varepsilon(Y_t^\varepsilon, \xi_t^\varepsilon)$ is replaced by the random quantity $\widetilde{\phi}_\varepsilon(t)$ defined by the *conditional shift*

$$\tilde{\phi}_\varepsilon(t) = -\frac{1}{\varepsilon} I\!\!E^* \left\{ \int_t^T (f(Y_s^\varepsilon, \xi_s^\varepsilon)^2 - \bar{\sigma}_\varepsilon^2) \, ds \mid \mathcal{F}_t^\varepsilon \right\}. \tag{10.40}$$

This is very similar to how solutions of the Poisson equations are constructed in the Markovian case since, after a change of variable $u = s/\varepsilon$, (10.40) becomes

$$\tilde{\phi}_\varepsilon(t) = -I\!\!E^* \left\{ \int_{t/\varepsilon}^{T/\varepsilon} (f(Y_u, \xi_u)^2 - \bar{\sigma}_\varepsilon^2) \, du \mid \mathcal{F}_{t/\varepsilon} \right\},$$

which is the non-Markovian equivalent of

$$\phi_\varepsilon(y) = -\int_0^{+\infty} I\!\!E^* \{ (f(Y_t)^2 - \bar{\sigma}_\varepsilon^2) \mid Y_0 = y \} \, dt,$$

a solution of the Poisson equation (10.27) as explained in Chapter 5 (see equation (5.25)). Exponential decay of the covariance function (Γ, for instance) and the properties of the OU process ensure that $\tilde{\phi}_\varepsilon(t)$ is well-defined by (10.40).

It is easily checked with the definition (10.40) that

$$\mathcal{M}_t^\varepsilon = \tilde{\phi}_\varepsilon(t) - \frac{1}{\varepsilon} \int_0^t (f(Y_s^\varepsilon, \xi_s^\varepsilon)^2 - \bar{\sigma}_\varepsilon^2) \, ds \tag{10.41}$$

is a martingale.

The second term in dN_t^ε given by (10.39) can be rewritten as

$$\frac{e^{-rt}}{2} (f(\tilde{Y}_t^\varepsilon)^2 - \bar{\sigma}_\varepsilon^2)(X_t^\varepsilon)^2 \frac{\partial^2 Q^\varepsilon}{\partial x^2}(t, X_t^\varepsilon) \, dt$$

$$= \frac{\varepsilon e^{-rt}}{2} (X_t^\varepsilon)^2 \frac{\partial^2 Q^\varepsilon}{\partial x^2}(t, X_t^\varepsilon) \{ d\tilde{\phi}_\varepsilon(t) - d\mathcal{M}_t^\varepsilon \}.$$

Repeating the argument given in Section 7.1.2, we perform the integration by parts to obtain

$$(X_t^\varepsilon)^2 \frac{\partial^2 Q^\varepsilon}{\partial x^2}(t, X_t^\varepsilon) \, d\tilde{\phi}_\varepsilon(t) = d\left((X_t^\varepsilon)^2 \frac{\partial^2 Q^\varepsilon}{\partial x^2}(t, X_t^\varepsilon) \tilde{\phi}_\varepsilon(t) \right)$$

$$- \tilde{\phi}_\varepsilon(t) \, d\left((X_t^\varepsilon)^2 \frac{\partial^2 Q^\varepsilon}{\partial x^2}(t, X_t^\varepsilon) \right)$$

$$- d\left\langle (X^\varepsilon)^2 \frac{\partial^2 Q^\varepsilon}{\partial x^2}(t, X^\varepsilon), \tilde{\phi}_\varepsilon \right\rangle_t,$$

where, from the martingale decomposition (10.41), the covariation term is given by

$$d\left\langle (X^\varepsilon)^2 \frac{\partial^2 Q^\varepsilon}{\partial x^2}(t, X^\varepsilon), \tilde{\phi}_\varepsilon \right\rangle_t = d\left\langle (X^\varepsilon)^2 \frac{\partial^2 Q^\varepsilon}{\partial x^2}(t, X^\varepsilon), \mathcal{M}^\varepsilon \right\rangle_t$$

$$= X_t^\varepsilon f(\tilde{Y}_t^\varepsilon) \frac{\partial}{\partial x}\left(x^2 \frac{\partial^2 Q^\varepsilon}{\partial x^2} \right)(t, X_t^\varepsilon) \, d\langle W^*, \mathcal{M}^\varepsilon \rangle_t.$$

We define $\psi_\varepsilon(t)$ by

$$\frac{\sqrt{\varepsilon}}{2} d\langle W^\star, \mathcal{M}^\varepsilon \rangle_t = \psi_\varepsilon(t)\, dt,$$

so that it plays the role of $(\nu\rho/\sqrt{2})\phi'(Y_t^\varepsilon)$ in the Markovian case. Putting all the nonmartingale terms of (10.39) together we see that, in order to center the $\sqrt{\varepsilon}$-order terms, $Q^\varepsilon = P_0^\varepsilon + \widetilde{Q}_1^\varepsilon$ should be chosen such that

$$\mathcal{L}_{\mathrm{BS}}(\bar{\sigma}_\varepsilon)Q^\varepsilon = \langle f\psi_\varepsilon \rangle_\varepsilon \frac{\partial}{\partial x}\left(x^2 \frac{\partial^2 P_0^\varepsilon}{\partial x^2}\right)$$

with the terminal condition $Q^\varepsilon(T, x) = P_0^\varepsilon(T, x) = h(x)$.

Defining the constant V_3^ε by

$$V_3^\varepsilon = \langle f\psi_\varepsilon \rangle_\varepsilon,$$

we conclude that Q^ε is characterized as in the Markovian case treated in Section 10.4.2, the difference being in the way the parameters $\bar{\sigma}_\varepsilon$ and V_3^ε are related to the model we started with. Applying the technique of expansion of the invariant distribution presented in Section 10.4.3, we deduce that a possible choice of approximated price is again given by $P_0 + \widetilde{P}_1$ as characterized in Chapter 5; the parameters $(\bar{\sigma}, V_2, V_3)$ are calibrated as summarized in Chapter 6.

10.6 Multidimensional Models

In this section we outline the extension of the asymptotic theory to the multidimensional case, where N underlying assets are driven by an N-dimensional Brownian motion with stochastic volatilities driven by an N-dimensional ergodic process.

Consider first the Black–Scholes case of constant volatilities. The asset prices (X_t) are geometric Brownian motions with the $N \times N$ invertible volatility matrix σ satisfying the stochastic differential equations

$$\frac{dX_i}{X_i} = \mu_i\, dt + \sum_{j=1}^{N} \sigma_{ij}\, dW_j,$$

where (W_t) is a standard Brownian motion in $I\!R^N$. We use subscripts now to denote components and omit the time dependence. In this complete market, there is a unique equivalent martingale measure $I\!P^\star$ under which

$$\frac{dX_i}{X_i} = r\, dt + \sum_{j=1}^{N} \sigma_{ij}\, dW_j^\star$$

for $i = 1, \ldots, N$ and $(W_t^\star)$ is a $I\!P^\star$-Brownian motion. Then a European contract with payoff function $h(X_T)$ has no-arbitrage price

$$P(t, x) = I\!E^\star \{e^{-r(T-t)} h(X_T) \mid X_t = x\}.$$

The pricing function $P(t, x)$ satisfies the PDE

$$\frac{\partial P}{\partial t} + \frac{1}{2} \sum_{i,j=1}^{N} v_{ij} x_i x_j \frac{\partial^2 P}{\partial x_i \partial x_j} + r \left(\sum_{j=1}^{N} x_j \frac{\partial P}{\partial x_j} - P \right) = 0,$$

where $\{v_{ij}\}$ is the symmetric diffusion matrix $\sigma \sigma^T$:

$$v_{ij} = \sum_{k=1}^{N} \sigma_{ik} \sigma_{jk}.$$

The terminal condition is $P(T, x) = h(x)$.

In other words, $P(t, x)$ is a function of the $N(N+1)/2$ independent entries of the matrix v, and consequently an implied volatility matrix is identified by $N(N+1)/2$ "independent" observed option prices.

Now consider the stochastic volatility extension:

$$\frac{dX_i}{X_i} = \mu_i \, dt + \sum_{j=1}^{N} f_{ij}(Y) \, dW_j, \tag{10.42}$$

$$dY_k = \alpha_k (m_k - Y_k) \, dt + \beta_k \, d\hat{Z}_k, \tag{10.43}$$

where $i, k = 1, \ldots, N$. The process $(\hat{Z}_t)$ is a standard Brownian motion in $I\!R^N$ that is correlated to (W_t) by

$$I\!E\{dW_j \, d\hat{Z}_k\} = d\langle W_j, \hat{Z}_k \rangle_t = \rho_{jk} \, dt.$$

Alternatively, we could write

$$\hat{Z}_k = \sum_{j=1}^{N} \rho_{jk} W_j + \left(1 - \sum_{j=1}^{N} \rho_{jk}^2 \right)^{1/2} Z_k,$$

where (W_t, Z_t) is a standard Brownian motion in $I\!R^{2N}$. Note that each volatility $f_{ij}(Y)$ is driven by the whole N-dimensional OU process (Y_t), so there is no need to incorporate further correlations between the components of $(\hat{Z}_t)$.

Under an equivalent martingale measure $I\!P^{\star(\Lambda)}$, we have

$$\frac{dX_i}{X_i} = r \, dt + \sum_{j=1}^{N} f_{ij}(Y) \, dW_j^\star, \tag{10.44}$$

$$dY_k = [\alpha_k (m_k - Y_k) - \beta_k \Lambda_k(Y)] \, dt + \beta_k \, d\hat{Z}_k^\star, \tag{10.45}$$

for some volatility risk premium $\Lambda(Y)$ chosen by the market and assumed to be a function of the OU process. A European contract has price given by

$$P(t, x, y) = I\!\!E^{\star(\Lambda)}\{e^{-r(T-t)}h(X_T) \mid X_t = x, \; Y_t = y\},$$

and the pricing function $P(t, x, y)$ satisfies

$$\frac{\partial P}{\partial t} + \frac{1}{2}\sum_{i,j=1}^{N} v_{ij} x_i x_j \frac{\partial^2 P}{\partial x_i \partial x_j} + \sum_{i,j,k=1}^{N} \rho_{jk}\beta_k f_{ij} x_i \frac{\partial^2 P}{\partial x_i \partial y_k}$$

$$+ \frac{1}{2}\sum_{k=1}^{N} \beta_k^2 \frac{\partial^2 P}{\partial y_k^2} + \frac{1}{2}\sum_{k\neq l} \beta_k \beta_l \left(\sum_j \rho_{jk}\rho_{jl}\right) \frac{\partial^2 P}{\partial y_k \partial y_l}$$

$$+ \sum_{k=1}^{N} [\alpha_k(m_k - y_k) - \beta_k \Lambda_k]\frac{\partial P}{\partial y_k} + r\left(\sum_{j=1}^{N} x_j \frac{\partial P}{\partial x_j} - P\right) = 0$$

with terminal condition $P(T, x, y) = h(x)$.

Introducing the usual scaling to model fast mean reversion in the volatilities, we write

$$\alpha_k = c_k/\varepsilon \quad \text{and} \quad \beta_k^2/2\alpha_k = v_k^2,$$

so that $\beta_k = v_k\sqrt{2c_k/\varepsilon}$. Note that each component of the OU process can mean revert at a different rate α_k but, with c_k fixed $\mathcal{O}(1)$ constants, the rates are assumed to be of the same order. The c_k are defined up to a multiplicative constant, which we can normalize by choosing (say) $c_1 = 1$. It is possible to introduce phenomena on different scales by allowing the c_k to be powers of ε, but this would lead to a more involved theory than we present here.

We now write our pricing problems as

$$\left(\frac{1}{\varepsilon}\mathcal{L}_0 + \frac{1}{\sqrt{\varepsilon}}\mathcal{L}_1 + \mathcal{L}_2\right)P^\varepsilon = 0,$$

where

$$\mathcal{L}_0 = \sum_k c_k \left(v_k^2 \frac{\partial^2}{\partial y_k^2} + (m_k - y_k)\frac{\partial}{\partial y_k}\right)$$

$$+ \sum_{k\neq l} v_k v_l \sqrt{c_k c_l}\left(\sum_j \rho_{jk}\rho_{jl}\right)\frac{\partial^2}{\partial y_k \partial y_l}, \qquad (10.46)$$

$$\mathcal{L}_1 = \sum_{i,j,k} v_k \sqrt{2c_k}\rho_{jk} f_{ij} x_i \frac{\partial^2}{\partial x_i \partial y_k} - \sum_k v_k \sqrt{2c_k}\Lambda_k \frac{\partial}{\partial y_k}, \qquad (10.47)$$

$$\mathcal{L}_2 = \frac{\partial}{\partial t} + \frac{1}{2}\sum_{i,j=1}^{N} v_{ij} x_i x_j \frac{\partial^2}{\partial x_i \partial x_j} + r\left(\sum_{j=1}^{N} x_j \frac{\partial}{\partial x_j} - \cdot\right). \qquad (10.48)$$

Notice that $\mathcal{L}_0$ is the infinitesimal generator of the N-dimensional OU process with the $\mathcal{N}(\boldsymbol{m}, \boldsymbol{v}^2)$-invariant distribution, where $\boldsymbol{m} \in \mathbb{R}^N$ has components m_k and $\boldsymbol{v}^2$ is the covariance matrix with diagonal entries v_k^2 and off-diagonal entries

$$\mathrm{cov}(Y_k, Y_l) = \frac{\beta_k \beta_l}{\alpha_k + \alpha_l} \sum_j \rho_{jk} \rho_{jl}.$$

We know from Chapter 5 that the key to constructing an asymptotic expansion

$$P^\varepsilon = P_0 + \sqrt{\varepsilon} P_1 + \varepsilon P_2 + \cdots$$

is the Poisson equation

$$\mathcal{L}_0 \chi(\boldsymbol{y}) + g(\boldsymbol{y}) = 0.$$

This has well-behaved solutions only if g is centered with respect to the invariant distribution of the N-dimensional OU process:

$$\langle g \rangle = \int_{R^N} \Phi(\boldsymbol{y}) g(\boldsymbol{y}) \, d\boldsymbol{y} = 0,$$

where $\Phi(\boldsymbol{y})$ is the density function for the N-dimensional normal distribution $\mathcal{N}(\boldsymbol{m}, \boldsymbol{v}^2)$ satisfying

$$\mathcal{L}_0^\star \Phi = 0.$$

Notice that, when $g = 0$, the solutions are constants in $\boldsymbol{y}$.

Then the arguments of Section 5.2 go through analogously and P_0 is a function, of $(t, \boldsymbol{x})$ only, that satisfies

$$\mathcal{L}_{\mathrm{BS}}^N(\bar{v}) P_0 = 0$$

with $P(T, \boldsymbol{x}) = h(\boldsymbol{x})$ and where we define $\mathcal{L}_{\mathrm{BS}}^N(v)$ by

$$\mathcal{L}_{\mathrm{BS}}^N = \frac{\partial}{\partial t} + \frac{1}{2} \sum_{i,j=1}^N v_{ij} x_i x_j \frac{\partial^2}{\partial x_i \partial x_j} + r \left(\sum_{j=1}^N x_j \frac{\partial}{\partial x_j} - \cdot \right),$$

a function of the symmetric matrix v. In this case an *effective volatility square matrix* $\bar{\sigma}$ satisfies $\bar{\sigma} \bar{\sigma}^T = \bar{v}$, with the average diffusion matrix given by

$$\bar{v}_{ij} = \sum_k \langle f_{ik} f_{jk} \rangle.$$

Continuing, we have that $\widetilde{P}_1 = \sqrt{\varepsilon} P_1$ is also a function of $(t, \boldsymbol{x})$ only and satisfies

$$\mathcal{L}_{\mathrm{BS}}^N(\bar{v}) \widetilde{P}_1 = \sqrt{\varepsilon} \langle \mathcal{L}_1 \mathcal{L}_0^{-1} (\mathcal{L}_2 - \langle \mathcal{L}_2 \rangle) \rangle P_0$$

with zero terminal condition. It remains to compute the operator $\mathcal{A}$ on the right-hand side.

Introducing the symmetric matrix $\phi(Y)$ satisfying the Poisson equations

$$\mathcal{L}_0\phi_{ij} = \sum_k f_{ik}f_{jk} - \bar{v}_{ij},$$

we generalize (5.45) as follows:

$$\mathcal{A} = \sum_{i,l,m} V_{ilm}^{(3)} x_i x_l x_m \frac{\partial^3}{\partial x_i \partial x_l \partial x_m} + \sum_{i,m} V_{im}^{(2)} x_i x_m \frac{\partial^2}{\partial x_i \partial x_m}.$$

Analogously to (5.39) and (5.40), the small constants $V_{im}^{(2)}$ and $V_{ilm}^{(3)}$ are given by

$$V_{im}^{(2)} = \sum_k \frac{v_k}{\sqrt{2\alpha}} \left(2 \sum_j \rho_{jk} \left\langle f_{ij} \frac{\partial \phi_{im}}{\partial y_k} \right\rangle - \left\langle \Lambda_k \frac{\partial \phi_{im}}{\partial y_k} \right\rangle \right), \qquad (10.49)$$

$$V_{ilm}^{(3)} = \sum_k \frac{v_k}{\sqrt{2\alpha}} \sum_j \rho_{jk} \left\langle f_{ij} \frac{\partial \phi_{lm}}{\partial y_k} \right\rangle, \qquad (10.50)$$

where $\alpha = \alpha_1$.

Notice that $V^{(3)}$ is symmetric in the last two indices. Hence, the parameter reduction is from N^2 unspecified functions f_{ij} and N volatility risk premia Λ_j plus $N(N+1)/2$ entries ρ_{ij} and $3N$ Y parameters (α, m, β) to $N^2(N+1)/2$ $V^{(3)}$ parameters plus N^2 $V^{(2)}$ parameters and $N(N+1)/2$ $\bar{v}$ entries.

Notes

The analysis of the constant volatility optimal asset allocation problem in Section 10.1.1 appears in Merton (1969, 1971). Much work has been done on these problems; see for example Duffie (1996) and Karatzas and Shreve (1998). Details about HJB equations and stochastic control can be found in Fleming and Soner (1993) and elsewhere.

The analysis of the stochastic volatility Merton problem in Section 10.1.2 is work in progress with Wendell Fleming. The distortion power transformation discussed at the end of Section 10.1.2 appears in Zariphopoulou (1999).

Averaging simultaneously with respect to random and periodic components has been studied extensively in the context of waves in random media; see Asch, Kohler, Papanicolaou, Postel, and White (1991). A reference for dealing with the general Markov processes of Section 10.3 is Ikeda and Watanabe (1989). The method of averaging by conditional shifts in non-Markovian situations is described in detail in Kushner (1984).

11 Applications to Interest-Rate Models

Our goal in this chapter is to show how the asymptotic method developed in this book in the context of the Black–Scholes theory is applied to interest rates. We do this by considering simple models of short rates (such as the Vasicek model) and computing corrections that come from a fast mean-reverting stochastic volatility. We show how these small corrections can greatly affect the shape of the term structure of interest rates, giving a simple and efficient tool for calibrating this structure (an analog of the smile curve).

11.1 Bond Pricing in the Vasicek Model

We apply our method to the simplest model for interest rates, the Vasicek model. It belongs to the family of affine models for which the yield curve is affine in the current short rate. This model permits explicit computations, and we show how to correct it for stochastic volatility. We do that first on bond prices and on the yield curve. In Section 11.3 we explain how to include stochastic volatility effects into other models such as the Cox–Ingersol–Ross (CIR) model.

11.1.1 Review of the Constant Volatility Vasicek Model

In the Vasicek model, the short rate is modeled as a mean-reverting Gaussian stochastic process $(r_t)_{t \geq 0}$ on a probability space $(\Omega, \mathcal{F}, I\!P)$ equipped with an increasing filtration $(\mathcal{F}_t)_{t \geq 0}$. Under the subjective probability measure $I\!P$, this process satisfies the linear stochastic differential equation (SDE)

$$dr_t = \hat{a}(\bar{r}_\infty - r_t)\, dt + \bar{\sigma}\, d\bar{W}_t, \qquad (11.1)$$

where $(\bar{W}_t)_{t \geq 0}$ is a standard Brownian motion. Here $\bar{\sigma}$ is its constant volatility.

Under an equivalent martingale (pricing) measure $I\!P^*$, it also follows a linear SDE

$$dr_t = \hat{a}(r^* - r_t)\, dt + \bar{\sigma}\, d\bar{W}_t^*, \tag{11.2}$$

where $(\bar{W}_t^*)$ is a standard $I\!P^*$-Brownian motion, if we assume the market price of interest-rate risk (denoted by λ) to be constant; λ is included in $r^* = \bar{r}_\infty - \lambda\bar{\sigma}/\hat{a}$. In other words, in the risk-neutral world $I\!P^*$, the short rate (r_t) is an Ornstein–Uhlenbeck process fluctuating around its mean level r^* with a rate of mean reversion $\hat{a}$. This process has the undesirable feature that it can become negative, but it is the simplest model in a larger family of models known as *affine*, for which the computation of bond prices and derivatives is relatively easy.

The no-arbitrage price at time t of a zero-coupon bond maturing at time T is given by

$$\mathcal{B}(t, T) = I\!E^* \left\{ \exp\left(-\int_t^T r_s\, ds\right) \mid \mathcal{F}_t \right\}. \tag{11.3}$$

Its discounted value is the martingale given by

$$\exp\left(-\int_0^t r_s\, ds\right)\mathcal{B}(t, T) = I\!E^* \left\{ \exp\left(-\int_0^T r_s\, ds\right) \mid \mathcal{F}_t \right\}.$$

Using the Markov property as in Section 1.4.4, the bond price $\mathcal{B}(t, T)$ is a function of the current short rate r_t. We denote this function by $P(t, x; T)$ to stress dependence on the current rate $r_t = x$. We have

$$\mathcal{B}(t, T) = P(t, r_t; T),$$

and according to the Feynman–Kac formula (1.67) (applied with $X_t = r_t$, the discount factor $\exp(-\int_t^T r_s\, ds)$, and the payoff $h = 1$), $P(t, x; T)$ is also the solution of the partial differential equation

$$\frac{\partial P}{\partial t} + \frac{1}{2}\bar{\sigma}^2\frac{\partial^2 P}{\partial x^2} + \hat{a}(r^* - x)\frac{\partial P}{\partial x} - xP = 0 \tag{11.4}$$

with the terminal condition $P(T, x; T) = 1$.

Introducing the (deterministic) time to maturity $\tau = T - t$, one can find a solution of the form

$$P(T - \tau, x; T) = A(\tau)e^{-B(\tau)x} \tag{11.5}$$

with $A(0) = 1$ and $B(0) = 0$; it is given by solving the ordinary differential equations

$$-B' = \hat{a}B - 1, \tag{11.6}$$

$$\frac{A'}{A} = \frac{1}{2}\bar{\sigma}^2 B^2 - \hat{a}r^* B, \tag{11.7}$$

which give

$$B(\tau) = \frac{1 - e^{-\hat{a}\tau}}{\hat{a}},$$ (11.8)

$$A(\tau) = \exp\left\{-\left[R_\infty \tau - R_\infty \frac{1 - e^{-\hat{a}\tau}}{\hat{a}} + \frac{\bar{\sigma}^2}{4\hat{a}^3}(1 - e^{-\hat{a}\tau})^2\right]\right\},$$ (11.9)

where we have set

$$R_\infty = r^* - \frac{\bar{\sigma}^2}{2\hat{a}^2}.$$

This leads to an explicit formula for the zero-coupon bond price $\mathcal{B}(t, T) = P(t, r_t; T)$:

$$\mathcal{B}(t, T) = \exp\left\{-\left[R_\infty(T - t) - (R_\infty - r_t)\frac{1 - e^{-\hat{a}(T-t)}}{\hat{a}} + \frac{\bar{\sigma}^2}{4\hat{a}^3}(1 - e^{-\hat{a}(T-t)})^2\right]\right\},$$ (11.10)

which could have been obtained also by solving the stochastic equation (11.2), as in Section 2.3.1:

$$r_t = r_0 e^{-\hat{a}t} + r^*(1 - e^{-\hat{a}t}) + \bar{\sigma}\int_0^t e^{-\hat{a}(t-s)} dW_s^*.$$ (11.11)

Its integral is

$$\int_t^T r_s\, ds = r^*(T - t) + (r^* - r_t)\frac{1 - e^{-\hat{a}(T-t)}}{\hat{a}} + \bar{\sigma}\int_t^T \frac{1 - e^{-\hat{a}(T-s)}}{\hat{a}} dW_s^*,$$ (11.12)

which, conditional on r_t, is normally distributed with mean

$$\mathbb{E}^*\left\{\int_t^T r_s\, ds \mid r_t\right\} = r^*(T - t) + (r^* - r_t)\frac{1 - e^{-\hat{a}(T-t)}}{\hat{a}}$$

and variance

$$\mathrm{var}^*\left\{\int_t^T r_s\, ds \mid r_t\right\} = \bar{\sigma}^2\int_t^T \left(\frac{1 - e^{-\hat{a}(T-s)}}{\hat{a}}\right)^2 ds.$$

Using the moment generating function of a normal random variable, we compute the expectation in (11.3) as

$$\mathbb{E}^*\left\{\exp\left(-\int_t^T r_s\, ds\right) \mid r_t\right\}$$
$$= \exp\left(-\mathbb{E}^*\left\{\int_t^T r_s\, ds \mid r_t\right\} + \frac{1}{2}\mathrm{var}^*\left\{\int_t^T r_s\, ds \mid r_t\right\}\right),$$

and we derive (11.10) using the explicit formulas for the conditional mean and variance.

Applying Itô's formula (1.16) to $\mathcal{B}(t, T)$ given by (11.10), we deduce

$$d\mathcal{B}(t, T) = \mathcal{B}(t, T)\left(r_t\, dt + \bar{\sigma}\,\frac{1 - e^{-\hat{a}(T-t)}}{\hat{a}}\, dW_t^*\right);$$

this shows that the *bond price volatility* is given by $(\bar{\sigma}/\hat{a})(1 - e^{-\hat{a}(T-t)})$, which is independent of r^*.

The *yield curve* is defined by

$$R(t, \tau) = -\frac{1}{\tau}\log(\mathcal{B}(t, t + \tau))$$

as a function of τ, and we deduce the explicit formula

$$R(t, \tau) = R_\infty - (R_\infty - r_t)\frac{1 - e^{-\hat{a}\tau}}{\hat{a}\tau} + \frac{\bar{\sigma}^2}{4\hat{a}^3\tau}(1 - e^{-\hat{a}\tau})^2,$$

which shows that $R(t, \tau)$ is an affine function of the current rate r_t and so justifies the name *affine model of term structure*. Observe that, for all t, $R(t, \tau)$ converges to $R_\infty = r^* - \bar{\sigma}^2/2\hat{a}^2$ as $\tau \to +\infty$.

Figure 11.1 shows a typical bond price evolution and the corresponding yield curve obtained under a Vasicek model. The yield curve (bottom) is a standard increasing curve resulting from this model when the current short rate is lower than the long-term rate.

11.1.2 Stochastic Volatility Vasicek Models

As we did for stock prices in Section 2.3.1, we introduce in equation (11.2) a stochastic volatility σ_t given by a nonnegative function $f(Y_t)$ of an Ornstein–Uhlenbeck process (Y_t). In the real world $\mathbb{P}$, the processes (r_t, Y_t) satisfy

$$dr_t = \hat{a}(\bar{r}_\infty - r_t)\, dt + f(Y_t)\, dW_t,$$

$$dY_t = \alpha(m - Y_t)\, dt + \beta\left(\rho\, dW_t + \sqrt{1 - \rho^2}\, dZ_t\right),$$

where (W_t) and (Z_t) are independent standard Brownian motions. The parameter ρ with $|\rho| < 1$ allows a correlation between the Brownian motion (W_t) driving the short rate and its volatility. Typically, we would expect $\rho > 0$ because rising volatility tends to push bond prices down and yields up. This is confirmed empirically.

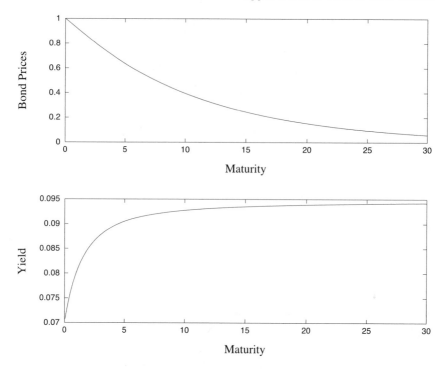

Figure 11.1. Bond prices (top) and yield curve (bottom) in the Vasicek model with $\hat{a} = 1$, $r^\star = 0.1$, and $\bar{\sigma} = 0.1$. Maturity τ runs from 0 to 30 years; $R_\infty = 0.095$ and the initial rate is $x = 0.07$.

In the risk-neutral world $I\!P^{\star(\lambda,\gamma)}$, the model for the short rate (r_t) becomes

$$dr_t = (\hat{a}(\bar{r}_\infty - r_t) - \lambda(Y_t) f(Y_t)) \, dt + f(Y_t) \, dW_t^\star, \qquad (11.13)$$

$$dY_t = \big(\alpha(m - Y_t) - \beta[\rho\lambda(Y_t) - \gamma(Y_t)\sqrt{1 - \rho^2}]\big) \, dt$$
$$+ \beta\big(\rho \, dW_t^\star + \sqrt{1 - \rho^2} \, dZ_t^\star\big), \qquad (11.14)$$

where $(W_t^\star)$ and $(Z_t^\star)$ are two independent standard $I\!P^{\star(\lambda,\gamma)}$-Brownian motions.

The market price of risk $\lambda(Y_t)$ may depend on Y_t, but we assume that it does not depend on the short rate r_t. Similarly, the *market price of volatility risk* $\gamma(Y_t)$ associated with the second source of randomness (Z_t) may depend on Y_t or may simply be a constant. This preserves the Markovian structure of the model and the autonomy of the volatility driving process (Y_t), so that (11.13)–(11.14) remains a genuine stochastic volatility model. Our assumption means that the market does *not* price term-structure derivatives as if the volatility were significantly affected by the level of the short-rate, or non-Markovian phenomena.

For general functions f, these models do not belong to the affine family in that yields will not be affine in the second factor Y_t. However, they do remain affine in the short rate r_t. Explicit formulas for bond prices through ordinary differential equations are not available.

The no-arbitrage price $\mathcal{B}(t, T)$ of a zero-coupon bond maturing at time T is now given by

$$P(t, x, y; T) = \mathbb{E}^{\star(\lambda, \gamma)}\left\{\exp\left(-\int_t^T r_s \, ds\right) \mid r_t = x, \, Y_t = y\right\}, \quad (11.15)$$

where the expectation $\mathbb{E}^{\star(\lambda, \gamma)}$ is taken with respect to the distribution of the (r_t, Y_t) solution of (11.13)–(11.14) starting at time t from (x, y). Using a two-dimensional Feynman–Kac formula, we deduce that $P(t, x, y; T)$ is the solution of the partial differential equation

$$\frac{\partial P}{\partial t} + \frac{1}{2} f(y)^2 \frac{\partial^2 P}{\partial x^2} + (\hat{a}(\bar{r}_\infty - x) - \lambda(y)f(y))\frac{\partial P}{\partial x} - xP + \beta \rho f(y)\frac{\partial^2 P}{\partial x \partial y}$$

$$+ \frac{1}{2}\beta^2 \frac{\partial^2 P}{\partial y^2} + \left(\alpha(m - y) - \beta[\rho\lambda(y) - \gamma(y)\sqrt{1 - \rho^2}]\right)\frac{\partial P}{\partial y} = 0, \quad (11.16)$$

with the terminal condition $P(T, x, y; T) = 1$ for every x and y.

In the context of fast mean-reverting stochastic volatility introduced in Chapter 3, the rate of mean reversion α of the process (Y_t) driving the volatility is large and $\nu = \beta/\sqrt{2\alpha}$ remains of order 1. This will be seen in yield curve data, presented in Section 11.4.

We use again the notation of Chapter 5:

$$\varepsilon = \frac{1}{\alpha},$$

$$\beta = \frac{\sqrt{2}\nu}{\sqrt{\varepsilon}},$$

$$P^\varepsilon(t, x, y; T) = P(t, x, y; T), \quad (11.17)$$

$$\mathcal{L}^\varepsilon = \frac{1}{\varepsilon}\mathcal{L}_0 + \frac{1}{\sqrt{\varepsilon}}\mathcal{L}_1 + \mathcal{L}_2.$$

Here $\mathcal{L}_0$ is, as in Chapter 5, the infinitesimal generator of the mean-reverting OU process (Y_t) scaled by $1/\alpha$,

$$\mathcal{L}_0 = v^2 \frac{\partial^2}{\partial y^2} + (m - y) \frac{\partial}{\partial y},$$

but $\mathcal{L}_2$ is now the "Vasicek-type" operator with volatility $f(y)$ and long-run mean $\bar{r}_\infty - \lambda f(y)/\hat{a}$:

$$\mathcal{L}_2 = \frac{\partial}{\partial t} + \frac{1}{2} f(y)^2 \frac{\partial^2}{\partial x^2} + (\hat{a}(\bar{r}_\infty - x) - \lambda f(y)) \frac{\partial}{\partial x} - x \cdot .$$

The operator

$$\mathcal{L}_1 = \sqrt{2} v \rho f(y) \frac{\partial^2}{\partial x \partial y} - \sqrt{2} v \left(\lambda(y)\rho + \gamma(y)\sqrt{1 - \rho^2} \right) \frac{\partial}{\partial y}$$

contains the two market prices of risk as well as terms proportional to the skew ρ. Equation (11.16) can be written as

$$\mathcal{L}^\varepsilon P^\varepsilon = 0$$

with the terminal condition $P^\varepsilon(T, x, y; T) = 1$.

As in Chapter 5, we look for an asymptotic solution of the form

$$P^\varepsilon(t, x, y; T) = P_0(t, x, y; T) + \sqrt{\varepsilon} P_1(t, x, y; T) + \varepsilon P_2(t, x, y; T) + \cdots,$$

with the terminal conditions $P_0(T, x, y; T) = 1$ and $P_1(T, x, y; T) = 0$. The $\mathcal{O}(\varepsilon^{-1})$-term implies that $P_0 = P_0(t, x; T)$ does not depend on y. The $\mathcal{O}(\varepsilon^{-1/2})$-terms give the same conclusion for $P_1 = P_1(t, x; T)$.

The $\mathcal{O}(1)$-terms lead to a Poisson equation in P_2 whose solvability condition is $\langle \mathcal{L}_2 P_0 \rangle = 0$, where $\langle \cdot \rangle$ denotes averaging with respect to the invariant distribution of the OU process (Y_t) as in Section 3.2.3. In other words, P_0 is the solution of $\langle \mathcal{L}_2 \rangle P_0 = 0$, which is exactly the partial differential equation (11.4) with the coefficients

$$\bar{\sigma}^2 = \langle f^2 \rangle \quad \text{and} \quad r^\star = \bar{r}_\infty - \langle \lambda f \rangle / \hat{a} \tag{11.18}$$

and with the same terminal condition $P_0(T, x; T) = 1$. The change of variable $\tau = T - t$ is again convenient, and we obtain

$$P_0(T - \tau, x; T) = A(\tau) e^{-B(\tau)x}, \tag{11.19}$$

where $A(\tau)$ and $B(\tau)$ are given explicitly by formulas (11.9) and (11.8), respectively. Notice that P_0 is exactly the Vasicek one-factor bond pricing function $\bar{P}$, with the "averaged" parameters $(\hat{a}, r^\star, \bar{\sigma})$ related to the stochastic volatility model parameters in (11.13)–(11.14) by (11.18).

The $\mathcal{O}(\sqrt{\varepsilon})$-terms give a Poisson equation in P_3 whose solvability condition is

$$\langle \mathcal{L}_2 P_1 + \mathcal{L}_1 P_2 \rangle = 0. \tag{11.20}$$

From the previous Poisson equation for P_2, we deduce that

$$P_2 = -\mathcal{L}_0^{-1}(\mathcal{L}_2 - \langle \mathcal{L}_2 \rangle) P_0 + k(t, x),$$

where $k(t, x)$ does not depend on y. The condition (11.20) gives the following equation for $P_1(t, x; T)$:

$$\langle \mathcal{L}_2 \rangle P_1 = \langle \mathcal{L}_1 \mathcal{L}_0^{-1}(\mathcal{L}_2 - \langle \mathcal{L}_2 \rangle) \rangle P_0,$$

with a zero terminal condition at $t = T$. Since we are mainly interested in the first correction to P_0, we rewrite this equation in terms of $\widetilde{P}_1(t, x; T) = \sqrt{\varepsilon} P_1(t, x; T)$ as

$$\langle \mathcal{L}_2 \rangle \widetilde{P}_1 = \mathcal{A} P_0, \tag{11.21}$$

with a zero terminal condition and where the operator $\mathcal{A}$ is defined by

$$\mathcal{A} = \sqrt{\varepsilon} \langle \mathcal{L}_1 \mathcal{L}_0^{-1}(\mathcal{L}_2 - \langle \mathcal{L}_2 \rangle) \rangle.$$

Note that $\widetilde{P}_1$ is of the same order $\sqrt{\varepsilon}$ as the source term $\mathcal{A} P_0$ in (11.21).

The operator $\mathcal{L}_2 - \langle \mathcal{L}_2 \rangle$ is given by

$$\mathcal{L}_2 - \langle \mathcal{L}_2 \rangle = \frac{1}{2}(f(y)^2 - \langle f^2 \rangle) \frac{\partial^2}{\partial x^2} - (\lambda(y)f(y) - \langle \lambda f \rangle) \frac{\partial}{\partial x}.$$

We introduce the centered functions ϕ and ψ, which are solutions of

$$\mathcal{L}_0 \phi = f(y)^2 - \langle f^2 \rangle \quad \text{and} \quad \mathcal{L}_0 \psi = \lambda(y)f(y) - \langle \lambda f \rangle.$$

Then the operator $\mathcal{A}$ can be written as

$$\mathcal{A} = V_1 \frac{\partial}{\partial x} + V_2 \frac{\partial^2}{\partial x^2} + V_3 \frac{\partial^3}{\partial x^3}, \tag{11.22}$$

where

$$V_3 = \frac{\nu}{\sqrt{2\alpha}} \rho \langle f\phi' \rangle,$$

$$V_2 = -\frac{\nu}{\sqrt{2\alpha}} \left(\rho \langle \lambda \phi' \rangle + \sqrt{1 - \rho^2} \langle \gamma \phi' \rangle \right) - \nu \rho \sqrt{\frac{2}{\alpha}} \langle f\psi' \rangle,$$

$$V_1 = \nu \sqrt{\frac{2}{\alpha}} \left(\rho \langle \lambda \psi' \rangle + \sqrt{1 - \rho^2} \langle \gamma \psi' \rangle \right)$$

are complicated functions of the model parameters, including market prices of risk and the volatility function f. Notice that no skew ($\rho = 0$) implies $V_3 = 0$ and that a market price of risk such that λf is constant implies $V_1 = 0$.

Using the change of variable $\tau = T - t$ together with (11.19) and (11.22), equation (11.21) becomes

$$\frac{\partial \widetilde{P}_1}{\partial \tau} = \frac{1}{2}\bar{\sigma}^2 \frac{\partial^2 \widetilde{P}_1}{\partial x^2} + \hat{a}(r^\star - x)\frac{\partial \widetilde{P}_1}{\partial x} - x\widetilde{P}_1$$
$$+ A(\tau)e^{-B(\tau)x}(V_3 B(\tau)^3 - V_2 B(\tau)^2 + V_1 B(\tau)) \qquad (11.23)$$

with the initial condition $\widetilde{P}_1(T - 0, x; T) = 0$.

We now try to find a solution of the form

$$\widetilde{P}_1(T - \tau, x; T) = D(\tau)A(\tau)e^{-B(\tau)x},$$

which leads to the following equation for $D(\tau)$:

$$D' = V_3 B^3 - V_2 B^2 + V_1 B.$$

Here we have used the equation (11.7) satisfied by B that is discussed in Section 11.1.1. Since $D(0) = 0$ and $A(0) = 1$, we obtain an explicit expression for $D(\tau)$, written here as a function of $B(\tau) = (1 - e^{-\hat{a}\tau})/\hat{a}$:

$$D(\tau) = \frac{V_3}{\hat{a}^3}\left(\tau - B(\tau) - \frac{1}{2}\hat{a}B(\tau)^2 - \frac{1}{3}\hat{a}^2 B(\tau)^3\right)$$
$$- \frac{V_2}{\hat{a}^2}\left(\tau - B(\tau) - \frac{1}{2}\hat{a}B(\tau)^2\right) + \frac{V_1}{\hat{a}}(\tau - B(\tau)). \qquad (11.24)$$

The bond price with stochastic volatility correction is thus

$$P(T - \tau, x, y; T) \approx P_0(T - \tau, x; T) + \widetilde{P}_1(T - \tau, x; T)$$
$$= (1 + D(\tau))A(\tau)e^{-B(\tau)x}; \qquad (11.25)$$

here D is proportional to $1/\sqrt{\alpha}$, which is small when we have fast mean reversion.

In Figure 11.2 we show the effect of the small correction $\widetilde{P}_1$ on the bond prices (top) and on the corresponding yield curve (bottom). The effect on the yield curve is very important qualitatively as well as quantitatively, since the curve can now have a hump.

11.2   Bond Option Pricing

In this section we show how to apply the asymptotic method and derive the corrections to the classical pricing formulas for European bond options written on zero-coupon bonds. We first recall these formulas in the constant volatility case and then calculate the correction for fast mean-reverting stochastic volatility.

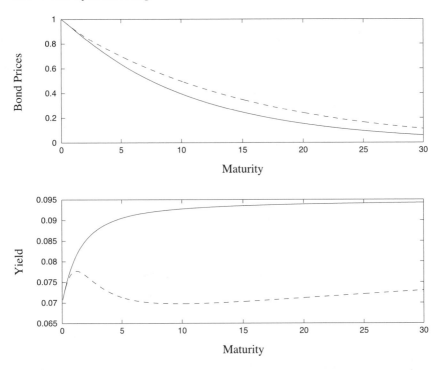

Figure 11.2. *Top.* Bond prices and corrected bond prices (dashed curve). *Bottom.* Yield curve and corrected yield curve (dashed) in the simulated Vasicek model (constant and stochastic volatility), with $\hat{a} = 1$, $r^{\star} = 0.1$, and $\bar{\sigma} = 0.1$ as in Figure 11.1. For the correction we have used $V_3 = 1/\sqrt{\alpha}$ assuming a nonzero skew, $\alpha = 10^3$, and $\lambda = \gamma = 0$ implying $V_1 = 0$ and $V_2 = 0$. Maturity τ runs from 0 to 30 years, and the initial rate is $x = 0.07$.

11.2.1   The Constant Volatility Case

We continue to denote the maturity of the bond by T and the maturity of a European option written on that bond by T_0, with $t \le T_0 < T$. As in the case of options on stocks (see Section 1.2.1), the payoff of the option at time T_0 is a function $h(\mathcal{B}(T_0, T))$ of the bond price at the expiration time of the option. We use here the notation $\mathcal{B}(t, T)$ introduced in Section 11.1.1 for bond prices. For instance, a call option is a contract that gives its owner the right but not the obligation to buy the bond at time T_0 at a predetermined strike price K. Its payoff, at maturity T_0, is then $h(\mathcal{B}(T_0, T)) = (\mathcal{B}(T_0, T) - K)^{+}$.

Under the Vasicek model (11.1), the no-arbitrage price of this option is obtained by computing the expectation of the discounted payoff with respect to the martingale measure $I\!\!P^{\star}$ for the bond. Recall that (r_t) is a Gaussian process satisfying the linear SDE (11.2) under $I\!\!P^{\star}$.

The payoff $h(\mathcal{B}(T_0, T))$ is a function $\tilde{h}(r_{T_0})$ of the rate at time T_0 because the bond price $\mathcal{B}(T_0, T)$ is a function of the rate r_{T_0}, as can be seen from the bond price formula (11.10). Using the Markovian structure as in Section 1.5, we denote by $Q(t, x; T, T_0)$ the price of this option at time t for an observed short rate $r_t = x$, and we compute it according to

$$Q(t, x) = I\!\!E^{\star}\left\{\exp\left(-\int_t^{T_0} r_s \, ds\right) h(\mathcal{B}(T_0, T)) \mid r_t = x\right\}$$

$$= I\!\!E^{\star}\left\{\exp\left(-\int_t^{T_0} r_s \, ds\right) \tilde{h}(r_{T_0}) \mid r_t = x\right\}. \qquad (11.26)$$

Using the Feynman–Kac formula (1.67) applied with $X_t = r_t$, the discount factor $\exp\left(-\int_t^{T_0} r_s \, ds\right)$, and the payoff $\tilde{h}(r_{T_0})$, we find that Q is also the solution of the partial differential equation

$$\frac{\partial Q}{\partial t} + \frac{1}{2}\bar{\sigma}^2 \frac{\partial^2 Q}{\partial x^2} + \hat{a}(r^{\star} - x)\frac{\partial Q}{\partial x} - xQ = 0,$$

with the terminal condition

$$Q(T_0, x) = \tilde{h}(x)$$

$$= h(\mathcal{B}(T_0, T))$$

$$= h\left(\exp\left(-\left(R_\infty(T_0 - t) - (R_\infty - x)B(T_0 - t) + \frac{\bar{\sigma}^2}{4\hat{a}}B(T_0 - t)^2\right)\right)\right),$$

where we recall from (11.8) and Section 11.1.1 that $B(T_0 - t) = (1 - e^{-\hat{a}(T_0 - t)})/\hat{a}$ and

$$R_\infty = r^{\star} - \frac{\bar{\sigma}^2}{2\hat{a}^2}.$$

Under $I\!\!P^{\star}$ and conditional on $r_t = x$, the pair of random variables $\left(r_{T_0}, \int_t^{T_0} r_s \, ds\right)$ is normally distributed according to (11.11) and (11.12). Therefore the price of the option given by (11.26) reduces to the computation of a Gaussian integral.

In the particular case of a call option, it is exercised if $\mathcal{B}(T_0, T) > K$, which can be written in terms of r_{T_0} as

$$r_{T_0} < r(T_0, K),$$

where the critical value $r(T_0, K)$ is defined by

$$r(T_0, K) = R_\infty \left(1 - \frac{T_0}{B(T_0)}\right) - \frac{\bar{\sigma}^2}{4\hat{a}} B(T_0) - \frac{\log K}{B(T_0)}.$$

Using the notation $C(t, x)$ for a call option, its price (11.26) can be rewritten as

$$C(t, x) = I\!\!E^* \left\{ \exp\left(-\int_t^{T_0} r_s\, ds\right) B(T_0, T) \mathbf{1}_{\{r_{T_0} < r(T_0, K)\}} \,\Big|\, r_t = x \right\}$$

$$- I\!\!E^* \left\{ \exp\left(-\int_t^{T_0} r_s\, ds\right) K \mathbf{1}_{\{r_{T_0} < r(T_0, K)\}} \,\Big|\, r_t = x \right\}$$

$$= B(t, T) I\!\!P_1\{r_{T_0} < r(T_0, K)\} - B(t, T_0) K I\!\!P_2\{r_{T_0} < r(T_0, K)\},$$

where $B(t, T)$ and $B(t, T_0)$ are evaluated at $r_t = x$ and, conditional on $\{r_t = x\}$, $I\!\!P_1$ and $I\!\!P_2$ have the following densities with respect to $I\!\!P^*$:

$$\frac{dI\!\!P_1}{dI\!\!P^*} = \frac{\exp\left(-\int_t^{T_0} r_s\, ds\right) B(T_0, T)}{B(t, T)} \quad \text{and} \quad \frac{dI\!\!P_2}{dI\!\!P^*} = \frac{\exp\left(-\int_t^{T_0} r_s\, ds\right)}{B(t, T_0)}.$$

The definition of $B(t, T_0)$ and the martingale property under $I\!\!P^*$ of the discounted bond prices $B(t, T)$ ensure that $I\!\!P_1$ and $I\!\!P_2$ are probabilities. One can show that, under $I\!\!P_1$ or $I\!\!P_2$, the random variable r_{T_0} is normally distributed and so we have

$$C(t, x) = B(t, T) N(h_1) - B(t, T_0) K N(h_2), \tag{11.27}$$

where N is the $\mathcal{N}(0, 1)$-distribution function (1.40). The quantities $h_{1,2}$ are given explicitly by

$$h_{1,2} = \frac{1}{v}\left(\log \frac{B(t, T)}{B(t, T_0)} - \log K \pm \frac{1}{2} v^2\right)$$

and

$$v^2 = \frac{\bar{\sigma}^2}{2\hat{a}^3}(1 - e^{-2\hat{a}(T_0 - t)})(1 - e^{-\hat{a}(T - t)})^2;$$

the bond prices $B(t, T)$ and $B(t, T_0)$ are given by (11.10) as functions of $r_t = x$.

11.2.2 *Correction for Stochastic Volatility*

We consider now the bond option pricing problem in the context of the stochastic volatility Vasicek model introduced in Section 11.1.2. The short rate process (r_t) and the OU process (Y_t) driving the volatility are given by equations (11.13) and (11.14). The option on the zero-coupon bond is the same as the one we just studied in the previous section. Its premium at time t is denoted by $Q(t, x, y; T, T_0)$. It is now a function of the current rate r_t and the current volatility level Y_t, given by

$$Q(t, x, y) = I\!\!E^{\star(\lambda, \gamma)} \left\{ \exp\left(-\int_t^{T_0} r_s \, ds\right) h(\mathcal{B}(T_0, T)) \mid r_t = x, \, Y_t = y \right\}, \quad (11.28)$$

where the bond price $\mathcal{B}(T_0, T)$ is a function of r_{T_0} and Y_{T_0} given by formula (11.15) applied with t replaced by T_0:

$$\mathcal{B}(T_0, T) = P(T_0, r_{T_0}, Y_{T_0}; T) = I\!\!E^{\star(\lambda, \gamma)} \left\{ \exp\left(-\int_{T_0}^T r_s \, ds\right) \mid r_{T_0}, Y_{T_0} \right\}.$$

By the Feynman–Kac formula, this premium $Q(t, x, y)$ is again a solution of the PDE (11.16), but the terminal condition at time T_0 becomes

$$Q(T_0, x, y) = h\left(I\!\!E^{\star(\lambda, \gamma)} \left\{ \exp\left(-\int_{T_0}^T r_s \, ds\right) \mid r_{T_0} = x, \, Y_{T_0} = y \right\} \right),$$

$$= h(P(T_0, x, y; T)), \quad (11.29)$$

where, as in (11.15), $P(T_0, x, y; T)$ denotes the bond price $\mathcal{B}(T_0, T)$ at time T_0 as a function of the short rate x and volatility level y.

If we compare what we have encountered so far with the pricing of stock options, this is a new situation in the sense that the terminal condition depends also on the volatility variable y. Fortunately, in the fast mean-reverting case (α large) we have already obtained, in Section 11.1.2, an expansion of the bond price $P(T_0, x, y; T)$ and we have seen that, up to order $\mathcal{O}(1/\sqrt{\alpha})$, this expansion is independent of y.

We keep here the notation (11.17) of Section 11.1.2: $\varepsilon = 1/\alpha$; the bond prices are denoted by $P^\varepsilon(t, x, y; T)$ and expanded as

$$P^\varepsilon(t, x, y; T) = P_0(t, x; T) + \sqrt{\varepsilon} P_1(t, x; T) + \cdots, \quad (11.30)$$

where the first two terms are given by (11.25), while the option prices will be denoted by $Q^\varepsilon(t, x, y)$ and expanded as

$$Q^\varepsilon(t, x, y) = Q_0(t, x, y) + \sqrt{\varepsilon} Q_1(t, x, y) + \cdots. \quad (11.31)$$

Assuming first that the payoff function h is smooth, one can expand the terminal condition (11.29) by using the first two terms in the bond-price expansion (11.25),

$$Q(T_0, x, y) = h(P_0(T_0, x; T)) + \sqrt{\varepsilon} P_1(T_0, x; T) h'(P_0(T_0, x; T)) + \cdots \quad (11.32)$$

and observing that, up to order $\mathcal{O}(\sqrt{\varepsilon})$, it does not depend on y.

Expanding the equation $\mathcal{L}^\varepsilon Q^\varepsilon = 0$, the term of order $1/\varepsilon$ gives that $Q_0(t, x, y)$ does not depend on y:

$$Q_0(t, x, y) = Q_0(t, x).$$

The terms of order $1/\sqrt{\varepsilon}$ lead to the same conclusion for $Q_1(t, x, y)$:

$$Q_1(t, x, y) = Q_1(t, x).$$

The $\mathcal{O}(1)$ terms lead to a Poisson equation in Q_2 whose solvability condition is $\langle \mathcal{L}_2 Q_0 \rangle = 0$, where $\langle \cdot \rangle$ denotes the averaging with respect to the invariant distribution of the OU process (Y_t). Since $Q_0(t, x)$ is independent of y, it is the solution of the partial differential equation

$$\langle \mathcal{L}_2 \rangle Q_0 = 0$$

with the terminal condition $Q_0(T_0, x) = h(P_0(T_0, x; T))$ at time T_0. This is exactly the problem solved in Section 11.2.1 in the case of a constant volatility $\bar{\sigma}$ and $r^\star = \bar{r}_\infty - \langle \lambda f \rangle / \hat{a}$. It was computed as a Gaussian integral in (11.26):

$$Q_0(t, x) = I\!E^\star \left\{ \exp\left(-\int_t^{T_0} r_s \, ds \right) h(P_0(T_0, x; T)) \right\} \tag{11.33}$$

$$= I\!E^\star \left\{ \exp\left(-\int_t^{T_0} r_s \, ds \right) h(A(T - T_0) e^{-B(T - T_0)}) \right\}, \tag{11.34}$$

where A and B are given by (11.9) and (11.8), respectively.

The $\mathcal{O}(\sqrt{\varepsilon})$ terms give a Poisson equation in Q_3 whose solvability condition, as in Section 11.1.2, reduces to

$$\langle \mathcal{L}_2 \rangle Q_1 = \mathcal{A} Q_0, \tag{11.35}$$

with the same operator

$$\mathcal{A} = V_3 \frac{\partial^3}{\partial x^3} + V_2 \frac{\partial^2}{\partial x^2} + V_1 \frac{\partial}{\partial x}$$

given by (11.22). Using (11.32), the terminal condition at time T_0 is now

$$Q_1(T_0, x) = P_1(T_0, x; T) h'(P_0(T_0, x; T)).$$

The solution to this problem has the following probabilistic representation:

$$Q_1(t, x) = I\!E^\star \left\{ \exp\left(-\int_t^{T_0} r_s \, ds \right) P_1(T_0, r_{T_0}; T) h'(P_0(T_0, r_{T_0}; T)) \right.$$
$$\left. - \int_t^{T_0} \exp\left(-\int_t^u r_s \, ds \right) \mathcal{A} Q_0(u, r_u) \, du \mid r_t = x \right\},$$

which reduces again to Gaussian integrals because, conditional on $r_t = x$, the pair $\left(\int_t^u r_s \, ds, r_u \right)$ is normally distributed for every $t \le u \le T_0$.

Using the representation (11.33) for the leading term $Q_0(t, x)$, we obtain the corrected bond option price formula as follows:

$$Q_0(t, x) + \frac{1}{\sqrt{\alpha}} Q_1(t, x)$$

$$= \mathbb{E}^\star \left\{ \exp\left(-\int_t^{T_0} r_s \, ds \right) \left[h(P_0(T_0, r_{T_0}; T)) \right. \right.$$

$$\left. \left. + \frac{1}{\sqrt{\alpha}} P_1(T_0, r_{T_0}; T) h'(P_0(T_0, r_{T_0}; T)) \right] \right\}$$

$$- \mathbb{E}^\star \left\{ \int_t^{T_0} \exp\left(-\int_t^u r_s \, ds \right) \frac{1}{\sqrt{\alpha}} \mathcal{A} Q_0(u, r_u) \, du \mid r_t = x \right\}.$$

Since the quantity

$$\left[h(P_0(T_0, x; T)) + \frac{1}{\sqrt{\alpha}} P_1(T_0, x; T) h'(P_0(T_0, x; T)) \right]$$

is independent of y and is the expansion, up to order $1/\sqrt{\alpha}$, of the terminal condition $h(P^\varepsilon(T_0, x, y; T))$, it is also equal, up to order $1/\sqrt{\alpha}$, to the average in the variable y of the terminal condition

$$\langle h(P^\varepsilon(T_0, x, y; T)) \rangle$$

with respect to the invariant distribution of OU process (Y_t). This last remark enables us to treat the case of nonsmooth payoff functions, since the integral with respect to y regularizes the derivatives of h. For instance, in the case of a European call bond option, the first term $Q_0(t, x)$ is equal to $C(t, x)$ obtained explicitly in the constant volatility case (11.27); the correction is

$$\frac{1}{\sqrt{\alpha}} Q_1(t, x)$$

$$= \mathbb{E}^\star \left\{ \exp\left(-\int_t^{T_0} r_s \, ds \right) \frac{1}{\sqrt{\alpha}} P_1(T_0, r_{T_0}; T) \mathbf{1}_{\{P_0(T_0, r_{T_0}; T) > K\}} \mid r_t = x \right\}$$

$$- \mathbb{E}^\star \left\{ \int_t^{T_0} \exp\left(-\int_t^u r_s \, ds \right) \frac{1}{\sqrt{\alpha}} \mathcal{A} C(u, r_u) \, du \mid r_t = x \right\}.$$

Table 11.1: *Parameters*

Model Parameters	Correction Parameters
Rate of mean reversion of short rate $\hat{a}$	$\hat{a}$
Long-run mean $\bar{r}_\infty$ under $I\!P$	$R^\star$
Specific volatility distribution $f(\cdot)$	Mean volatility $\bar{\sigma}$
Rate of mean-reversion of volatility α	Group parameter V_1
Mean level m of (Y_t)	Group parameter V_2
"V-vol" β	Group parameter V_3
Interest-rate risk premium $\lambda(\cdot)$	
Volatility risk premium $\gamma(\cdot)$	

Using the function D defined in (11.24) and the formula (11.27) for $C(t, x)$, the corrected call bond price becomes

$$C(t, x) + \frac{D(T - T_0)}{\sqrt{\alpha}} \mathcal{B}(t, T) N(h_1)$$

$$- \frac{1}{\sqrt{\alpha}} \int_t^{T_0} I\!E^\star \left\{ \exp\left(-\int_t^u r_s \, ds \right) \mathcal{A} C(u, r_u) \mid r_t = x \right\} du,$$

where $\mathcal{B}(t, T)$ is given by (11.10) and the operator $\mathcal{A}$ by (11.22).

11.2.3   Implications

The main implication of the asymptotic calculation is reduction of the parametric dependence of the pricing formulas on a specific model. In Table 11.1, the left column lists the original model parameters (including the risk premium functions), and the right column lists the group parameters that need to be estimated from the yield curve using formula (11.25).

In particular, the sensitivities to the correlation ρ and the rate of mean reversion α of the *hidden* process (Y_t) are absorbed in the group parameters V_1, V_2, V_3. There are also the added benefits that (a) the functions $f(\cdot)$, $\lambda(\cdot)$, and $\gamma(\cdot)$ need not be specified and (b) to this level of approximation, the present level y of the unobserved volatility driving process does not need to be estimated.

The stochastic volatility–corrected prices of bond options also depend, to this order, on these same parameters. That is, although the group parameters do not give us enough to recover the law of the processes (r_t, Y_t), they do give us enough to price other interest-rate derivatives under fast mean-reverting stochastic volatility.

11.3 Asymptotics around the CIR Model

In this section we present the fast mean-reverting stochastic volatility asymptotic calculation for a model built around the popular CIR model; this will illustrate the flexibility of the method under changes to the background short-rate model. Because we would like a closed-form correction for bond prices, we mimic the form of the two-factor CIR model:

$$dr_t = \kappa_1(\mu_1 - r_t)\,dt + \sqrt{c_{11}r_t + c_{12}V_t}\,dW_t^\star,$$

$$dV_t = \kappa_2(\mu_2 - V_t)\,dt + \sqrt{c_{21}r_t + c_{22}V_t}\,dZ_t^\star,$$

where the independent Brownian motions $(W_t^\star, Z_t^\star)$ are preceded by "square root of affine" coefficients. This means that the derivative pricing PDE has linear coefficients, which admit an affine yield curve solution.

We shall look at the model

$$dr_t = \hat{a}(r^\star - r_t)\,dt + f(Y_t)\sqrt{r_t}\,dW_t^\star,$$

$$dY_t = \alpha r_t(m - Y_t)\,dt + \beta\sqrt{r_t}\left(\rho\,dW_t^\star + \sqrt{1 - \rho^2}\,dZ_t^\star\right),$$

where we are already under the pricing measure. (We will not look at the link between the real measure and the risk-neutral measure for this model.) Notice that the second process (Y_t) driving the volatility is no longer autonomous in that it depends on (r_t). We make this change to admit yield curves that are affine in the short-rate level $x = r_t$ but not, of course, in $y = Y_t$. This allows a closed form (up to solution of ODEs) for the stochastic volatility correction. We assume that $f(y)^2 \leq 2\hat{a}r^\star$ for all y, which guarantees the existence of a strong solution for (r_t) that stays strictly positive.

The set-up for bond pricing is exactly as described in Section 11.1.2, with the new differential operators

$$\mathcal{L}_0 = x\left(v^2\frac{\partial^2}{\partial y^2} + (m - y)\frac{\partial}{\partial y}\right),$$

$$\mathcal{L}_1 = \sqrt{2}v\rho x f(y)\frac{\partial^2}{\partial x\partial y},$$

$$\mathcal{L}_2 = \frac{\partial}{\partial t} + \frac{1}{2}f(y)^2 x\frac{\partial^2}{\partial x^2} + \hat{a}(r^\star - x)\frac{\partial}{\partial x} - x\cdot.$$

Notice that $\mathcal{L}_0$ is merely x times the generator of the OU process (Y_t) scaled by $1/\alpha$ and that $\mathcal{L}_2$ is the CIR operator with volatility $f(y)$.

Then the arguments of Section 11.1.2 go through analogously, with the extra x-factor in $\mathcal{L}_0$ not affecting the conclusions following from the Poisson equations in y. Bond prices $P^\varepsilon(t, x, y)$ are approximated by

$$P^\varepsilon(T - \tau, x, y) \approx P_0(T - \tau, x) + \widetilde{P}_1(T - \tau, x),$$

where the first term solves

$$\langle \mathcal{L}_2 \rangle P_0 = \left(\frac{\partial}{\partial t} + \frac{1}{2}\bar{\sigma}^2 x \frac{\partial^2}{\partial x^2} + \hat{a}(r^\star - x)\frac{\partial}{\partial x} - x \right) P_0 = 0,$$

$$P_0(T, x) = 1,$$

and is given by the one-factor affine CIR formulas

$$P_0(T - \tau, x) = A(\tau)e^{-B(\tau)x},$$

$$A(\tau) = \left(\frac{2\theta e^{(\theta + \hat{a})\tau/2}}{(\theta + \hat{a})(e^{\theta\tau} - 1) + 2\theta} \right)^{2\hat{a}r^\star/\bar{\sigma}^2}, \tag{11.36}$$

$$B(\tau) = \frac{2(e^{\theta\tau} - 1)}{(\theta + \hat{a})(e^{\theta\tau} - 1) + 2\theta}, \tag{11.37}$$

$$\theta = \sqrt{\hat{a}^2 + 2\bar{\sigma}^2}.$$

The correction can be shown to satisfy the analog of (11.21), which (with the operators just defined) reduces to

$$\langle \mathcal{L}_2 \rangle \widetilde{P}_1 = -V_3 x \frac{\partial^3 P_0}{\partial x^3},$$

$$P_1(T, x) = 0,$$

where V_3 is some group parameter related to the original parameters that can be estimated from the yield curve. This equation is solved to give

$$\widetilde{P}_1(T - \tau, x) = (D_1(\tau)x + D_2(\tau))A(\tau)e^{-B(\tau)x},$$

where A and B are given by (11.36)–(11.37) and D_1 and D_2 satisfy the ODEs

$$D_1' = V_3 B^3 - (\bar{\sigma}^2 B + \hat{a})D_1,$$

$$D_2' = \hat{a}r^\star D_1,$$

with zero initial conditions.

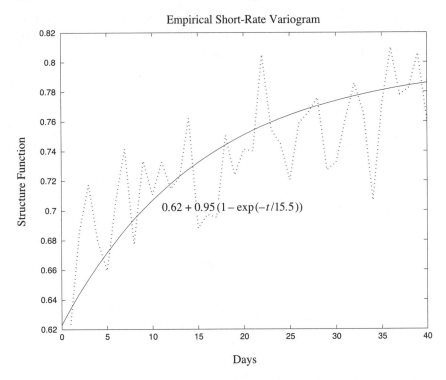

Figure 11.3. Variogram of the log of the absolute value of the differenced short rate. The correlation time is approximately two weeks.

11.4   Illustration from Data

We present some results from an analysis of yield curve data indicating fast mean reversion in the short-rate volatility, which justifies the asymptotic analysis. Then we show snapshot fits of the yield curve fitted to the asymptotic formula of Section 11.1.2, in comparison to a CIR fit. Description of the data and the full analysis appears in work referenced in the notes at the end of this chapter.

11.4.1   Variogram Analysis

We apply the variogram method of Chapter 4 to the log of absolute differences of yields of short maturity, which proxy for the short rate; this is analogous to (4.6). The exponential decay fit in Figure 11.3 of the data's structure function reveals a

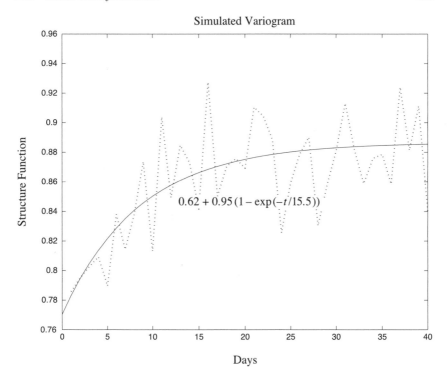

Figure 11.4. Variogram from simulated two-scale expOU stochastic volatility Vasicek model. The parameters used were $\hat{a} = 0.7$, $\bar{r}_\infty = 0.06$, $\alpha = 26$, $m = -5$, $\beta = 2$, $\rho = 0.6$, and $\lambda = \gamma = 0$.

characteristic time of mean reversion *in the short-rate volatility* of the order of two weeks. This is fast in comparison with the maturities of bonds, which are on the scale of years. As in Chapter 4, we tested the variogram method on simulated data using reasonable parameter values; this is shown in Figure 11.4.

11.4.2 Yield Curve Fitting

In order to illustrate how the parsimonious group parameters capture complicated yield curve shapes, such as humps, we fit the formula (11.25) for the yield curve on a given day during the bond crisis of 1998. The snapshot fit is shown in the top graph of Figure 11.5. For comparison, we show in the bottom graph the same snapshot fit using a single-factor CIR model with jumps.

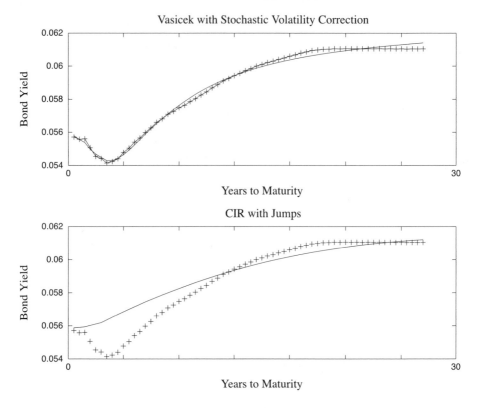

Figure 11.5. Snapshot of the yield curve fit with the stochastic volatility–corrected Vasicek model (top) and with the single-factor CIR model and down jumps (bottom) for September 6, 1998.

Notes

Stochastic models of the short rate are discussed extensively in Hull (1999), Duffie (1996), and Musiela and Rutkowski (1997), for instance. Empirical evidence of randomly changing short-rate volatility and stochastic volatility models are described in Longstaff and Schwartz (1992). Multifactor affine models are detailed in Duffie and Kan (1996).

The asymptotic and data analysis is part of a more extensive work (Cotton, Fouque, Papanicolaou, and Sircar 1999) in collaboration with Peter Cotton.

Bibliography

Asch, M., Kohler, W., Papanicolaou, G., Postel, M., and White, B. (1991). Frequency content of randomly scattered signals. *SIAM Review* 33(4): 519–625.

Avellaneda, M., Friedman, C., Holmes, R., and Samperi, D. (1997). Calibrating volatility surfaces via relative-entropy minimization. *Appl. Math. Finance* 4(1): 37–64.

Ball, C., and Roma, A. (1994). Stochastic volatility option pricing. *J. Financial and Quantitative Analysis* 29(4): 589–607.

Bensoussan, A., and Lions, J.-L. (1982). *Applications of Variational Inequalities in Stochastic Control* (Studies in Mathematics and its Applications, vol. 12). Amsterdam: North-Holland.

Bensoussan, A., Lions, J.-L., and Papanicolaou, G. (1978). *Asymptotic Analysis for Periodic Structures.* Amsterdam: North-Holland.

Black, F., and Scholes, M. (1973). The pricing of options and corporate liabilities. *J. Political Economy* 81: 637–59.

Blankenship, G., and Papanicolaou, G. C. (1978). Stability and control of stochastic systems with wide-band noise disturbances. *SIAM J. Appl. Math.* 34(3): 437–76.

Breiman, L. (1992). *Probability* (Classics in Applied Mathematics, vol. 7). Philadelphia: SIAM.

Cotton, P., Fouque, J.-P., Papanicolaou, G., and Sircar, K. R. (1999). Stochastic volatility corrections for interest rate models (preprint).

Derman, E., and Kani, I. (1994). Riding on a smile. *RISK* 7: 32–9.

Duffie, D. (1996). *Dynamic Asset Pricing Theory*, 2nd ed. Princeton, NJ: Princeton University Press.

Duffie, D., and Kan, R. (1996). A yield-factor model of interest rates. *Math. Finance* 6(4): 379–406.

Dumas, B., Fleming, J., and Whaley, R. (1998). Implied volatility functions: Empirical tests. *J. Finance* 53(6): 2059–2106.

Dupire, B. (1994). Pricing with a smile. *RISK* 7: 18–20.

El Karoui, N., and Quenez, M. (1995). Dynamic programming and pricing of contingent claims in an incomplete market. *SIAM J. Control and Optimization* 33: 29–66.

Fleming, W. H., and Soner, H. M. (1993). *Controlled Markov Processes and Viscosity Solutions.* Berlin: Springer-Verlag.

Fouque, J.-P., Papanicolaou, G., and Sircar, K. R. (1998). Asymptotics of a two-scale stochastic volatility model. In *Equations aux dérivées partielles et applications, in honour of Jacques-Louis Lions.* Paris: Gauthier-Villars, pp. 517–25.

Fouque, J.-P., Papanicolaou, G., and Sircar, K. R. (1999a). Financial modeling in a fast mean-reverting stochastic volatility environment. *Asia-Pacific Financial Markets* 6(1): 37–48.

Fouque, J.-P., Papanicolaou, G., and Sircar, K. R. (1999b). From the implied volatility skew to a robust correction to Black–Scholes American option prices. *International J. Theoretical and Appl. Finance* (to appear).

Fouque, J.-P., Papanicolaou, G., and Sircar, K. R. (2000a). Mean-reverting stochastic volatility. *International J. Theoretical and Appl. Finance* 3(1): 101–42.

Fouque, J.-P., Papanicolaou, G., and Sircar, K. R. (2000b). Random volatility. *RISK Magazine* 13(2): 89–92.

Fouque, J.-P., Papanicolaou, G., Sircar, K. R., and Solna, K. (1999). Mean reversion of S&P 500 volatility (preprint).

Frey, R. (1996). Derivative asset analysis in models with level-dependent and stochastic volatility. *CWI Quarterly* 10(1): 1–34.

Ghysels, E., Harvey, A., and Renault, E. (1996). Stochastic volatility. In G. Maddala and C. Rao (Eds.), *Statistical Methods in Finance* (Handbook of Statistics, vol. 14), pp. 119–91. Amsterdam: North Holland.

Heston, S. (1993). A closed-form solution for options with stochastic volatility with applications to bond and currency options. *Review of Financial Studies* 6(2): 327–43.

Hobson, D. (1996). Stochastic volatility. Technical Report, School of Mathematical Sciences, University of Bath, U.K.

Hull, J. (1999). *Options, Futures and Other Derivative Securities*, 4th ed. Englewood Cliffs, NJ: Prentice-Hall.

Hull, J., and White, A. (1987). The pricing of options on assets with stochastic volatilities. *J. Finance* 42(2): 281–300.

Ikeda, N., and Watanabe, S. (1989). *Stochastic Differential Equations and Diffusion Processes,* 2nd ed. Amsterdam: North-Holland/Kodansha.

Jackwerth, J., and Rubinstein, M. (1996). Recovering probability distributions from contemporaneous security prices. *J. Finance* 51(5): 1611–31.

Karatzas, I., and Shreve, S. (1998). *Methods of Mathematical Finance.* Berlin: Springer-Verlag.

Kushner, H. (1984). *Approximation and Weak Convergence Methods for Random Processes, with Applications to Stochastic Systems Theory* (MIT Press Series in Signal Processing, Optimization, and Control, no. 6). Cambridge, MA: MIT Press.

Lamberton, D., and Lapeyre, B. (1996). *Introduction to Stochastic Calculus Applied to Finance.* New York: Chapman & Hall.

Lee, R. (1999). Local volatilities under stochastic volatility. *International J. Theoretical and Appl. Finance* (to appear).

Longstaff, F., and Schwartz, E. (1992). Interest rate volatility and the term structure: A two-factor general equilibrium model. *J. Finance* 47(4): 1259–82.

Merton, R. C. (1969). Lifetime portfolio selection under uncertainty: The continuous-time case. *Review of Economic Statistics* 51: 247–57.

Merton, R. C. (1971). Optimum consumption and portfolio rules in a continuous-time model. *J. Economic Theory* 3(1/2): 373–413.

Musiela, M., and Rutkowski, M. (1997). *Martingale Methods in Financial Modelling.* Berlin: Springer-Verlag.

Oksendal, B. (1998). *Stochastic Differential Equations: An Introduction with Applications,* 5th ed. New York: Springer.

Papanicolaou, G. C. (1978). Asymptotic analysis of stochastic equations. In M. Rosenblatt (Ed.), *Studies in Probability Theory,* pp. 111–79. Washington, DC: Mathematical Association of America.

Papanicolaou, G. C., Stroock, D., and Varadhan, S. R. S. (1977). Martingale approach to some limit theorems. In M. Reed (Ed.), *Statistical Mechanics, Dynamical Systems and the Duke Turbulence Conference* (Duke University Mathematics Series, vol. 3). Durham, NC: Duke University Press.

Renault, E., and Touzi, N. (1996). Option hedging and implied volatilities in a stochastic volatility model. *Math. Finance* 6(3): 279–302.

Rubinstein, M. (1985). Nonparametric tests of alternative option pricing models. *J. Finance* 40(2): 455–80.

Rubinstein, M. (1994). Implied binomial trees. *J. Finance* 69(3): 771–818.

Samuelson, P. (1973). Mathematics of speculative prices. *SIAM Review* 15: 1–39.

Schweizer, M. (1999). A guided tour through quadratic hedging approaches. Preprint, Technische Universität, Berlin.

Scott, L. (1987). Option pricing when the variance changes randomly: Theory, estimation, and an application. *J. Financial and Quantitative Analysis* 22(4): 419–38.

Sircar, K. R. (1999). Hedging under stochastic volatility. In M. Avellaneda (Ed.), *Courant Finance Seminar,* vol. 2. Singapore: World Scientific.

Sircar, K. R., and Papanicolaou, G. C. (1999). Stochastic volatility, smile and asymptotics. *Appl. Math. Finance* 6(2): 107–45.

Stein, E., and Stein, J. (1991). Stock price distributions with stochastic volatility: An analytic approach. *Review of Financial Studies* 4(4): 727–52.

Wiggins, J. (1987). Option values under stochastic volatility. *J. Financial Economics* 19(2): 351–72.

Willard, G. (1996). Calculating prices and sensitivities for path-independent derivative securities in multifactor models. Ph.D. thesis, Washington University in St. Louis, MO.

Wilmott, P., Howison, S., and Dewynne, J. (1996). *Mathematics of Financial Derivatives: A Student Introduction.* Cambridge University Press.

Zariphopoulou, T. (1999). A solution approach to valuation with unhedgeable risks. *Finance and Stochastics* (to appear).

Index